Systems Architecture of Patient Surveillance Cloud Applications and Services IoT System

-- Structure-Behavior Coalescence Method for Model Singularity --

William S. Chao

Structure-Behavior Coalescence

Systems Architecture = Systems Structure + Systems Behavior

CONTENTS

CONTENTS 5
PREFACE 7
ABOUT THE AUTHOR 9
PART I: BASIC CONCEPTS 11
    Chapter 1: Introduction to Patient Surveillance Cloud Applications and Services IoT System 13
        1-1 Behavior of Registering_Patient_Account 16
        1-2 Behavior of Sensing_Patient_Vital_Signs 17
        1-3 Behavior of Alerts_Notifying 18
        1-4 Behavior of Recording_Emergency_Response_Starting_Time 19
        1-5 Behavior of Recording_Emergency_Response_End_Time 20
    Chapter 2: Introduction to General Systems Theory 2.0 23
        2-1 Multiple Views of a System 23
        2-2 Non-Architectural Approach versus Architectural Approach 26
        2-3 Definition of Systems Architecture 28
        2-4 Architecture Description Language 30
        2-5 Multiple Views Coalescence to Achieve the Systems Architecture 30
        2-6 Integrating the Systems Structures and Systems Behaviors 32
        2-7 Structure-Behavior Coalescence to Facilitate Multiple Views Coalescence 34
        2-8 Structure-Behavior Coalescence to Achieve the Systems Architecture 35
        2-9 Using SBC-ADL to Construct the Systems Architecture 36
        2-10 SBC Model Singularity 39
PART II: SBC ARCHITECTURE DESCRIPTION LANGUAGE .. 43
    Chapter 3: Systems Structure 45
        3-1 Architecture Hierarchy Diagram 45
            3-1-1 Decomposition and Composition 45
            3-1-2 Multi-Level Decomposition and Composition 48
            3-1-3 Aggregated and Non-Aggregated Systems 50
        3-2 Framework Diagram 51
            3-2-1 Multi-Layer Decomposition and Composition 51
            3-2-2 Only Non-Aggregated Systems Appearing in Framework Diagrams 53
        3-3 Component Channel Diagram 54
            3-3-1 Channels of Components 54

   3-3-2 Drawing the Component Channel Diagram 58
  3-4 Component Connection Diagram 62
   3-4-1 Essence of a Connection..62
   3-4-2 Drawing the Component Connection Diagram 64

Chapter 4: Systems Behavior..67
  4-1 Structure-Behavior Coalescence Diagram...67
   4-1-1 Purpose of Structure-Behavior Coalescence Diagram 67
   4-1-2 Drawing the Structure-Behavior Coalescence Diagram.............68
  4-2 Interaction Flow Diagram 70
   4-2-1 Individual Systems Behavior Represented by Interaction Flow Diagram..70
   4-2-2 Drawing the Interaction Flow Diagram......................................71

PART III: SYSTEMS ARCHITECTURE OF PSCASIS 75
 Chapter 5: AHD of the PSCASIS 77
 Chapter 6: FD of the PSCASIS..81
 Chapter 7: CChD of the PSCASIS 83
 Chapter 8: CCoD of the PSCASIS 97
 Chapter 9: SBCD of the PSCASIS 99
 Chapter 10: IFD of the PSCASIS 103

APPENDIX A: SBC ARCHITECTURE DESCRIPTION LANGUAGE...111
APPENDIX B: SBC PROCESS ALGEBRA 117
BIBLIOGRAPHY 121

PREFACE

A system is complex that it comprises multiple views such as strategy/version n, strategy/version n+1, concept, analysis, design, implementation, structure, behavior and input/output data views. Accordingly, a system is defined as a set of interacting components forming an integrated whole of that system's multiple views.

Since structure and behavior views are the two most prominent ones among multiple views, integrating the structure and behavior views is a method for integrating multiple views of a system. In other words, structure-behavior coalescence (SBC) is a single model (model singularity) approach which results in the integration of multiple views. Therefore, it is concluded that the SBC architecture is so proper to model the multiple views of a system.

In this book, we use the SBC architecture description language (SBC-ADL) to describe and represent the systems architecture of Patient Surveillance Cloud Applications and Services IoT System (PSCASIS). An architecture description language is a special kind of system model used in defining the architecture of a system. SBC-ADL uses six fundamental diagrams to formally grasp the essence of a system and its details at the same time. These diagrams are: a) architecture hierarchy diagram, b) framework diagram, c) component channel diagram, d) component connection diagram, e) structure-behavior coalescence diagram and f) interaction flow diagram.

Systems architecture is on the rise. By this book's introduction and elaboration of the systems architecture of PSCASIS, all readers may understand clearly how the SBC-ADL helps architects effectively perform architecting, in order to synergistically construct the fruitful systems architecture.

8

ABOUT THE AUTHOR

 Dr. William S. Chao is the CEO & founder of SBC Architecture International®. SBC (Structure-Behavior Coalescence) architecture is a systems architecture which demands the integration of systems structure and systems behavior of a system. SBC architecture applies to hardware architecture, software architecture, enterprise architecture, knowledge architecture and thinking architecture. The core theme of SBC architecture is: Architecture = Structure -->> Behavior.

William S. Chao received his bachelor degree (1976) in telecommunication engineering and master degree (1981) in information engineering, both from the National Chiao-Tung University, Taiwan. From 1976 till 1983, he worked as an engineer at Chung-Hwa Telecommunication Company, Taiwan.

William S. Chao received his master degree (1985) in information science and Ph.D. degree (1988) in information science, both from the University of Alabama at Birmingham, USA. From 1988 till 1991, he worked as a computer scientist at GE Research and Development Center, Schenectady, New York, USA.

Dr. William S. Chao has been teaching at National Sun Yat-Sen University, Taiwan since 1992 and now serves as the president of Association of Enterprise Architects, Taiwan Chapter. His research covers: systems architecture, hardware architecture, software architecture, enterprise architecture, knowledge architecture and thinking architecture.

10

PART I: BASIC CONCEPTS

Chapter 1: Introduction to Patient Surveillance Cloud Applications and Services IoT System

Continuous measurement of patient parameters such as heart rate and rhythm, respiration rate, blood pressure, oxygen saturation and many other parameters has become a common feature of the care of critically ill patients. When accurate and immediate decision-making is crucial for effective patient care, electronic monitors frequently are used to collect and display physiological data.

Patient surveillance has been rigorously defined by Philip Jevon [Jevo12] as "repeated or continuous observations or measurements of the patient, his or her physiological function, and the function of life support equipment, for the purpose of guiding management decisions, including when to make therapeutic interventions, and assessment of those interventions."

Patient surveillance cloud applications and services IoT system (PSCASIS) provides real-time patient monitoring solution as shown in Figure 1-1. PSCASIS may not only alert doctors or nurses to potentially life-threatening events, it also provides physiologic input data used to control directly connected life-support devices.

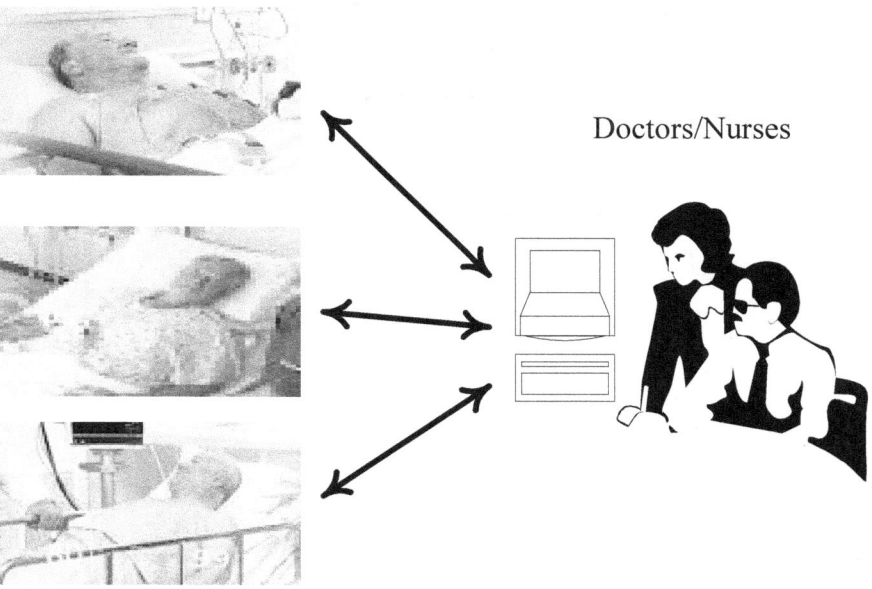

Figure 1-1 PSCASIS Provides Real-Time Patient Monitoring Solution

A cloud applications and services Internet of Things (IoT) system is the network of physical things or objects embedded with electronics, software, sensors and connectivity to enable it to achieve greater value and service by exchanging data with the users, operators and other connected devices. Each thing or object is uniquely identifiable through its embedded cloud computing [Bern09] system but is able to interoperate within the existing internet infrastructure, as shown in Figure 1-2.

Figure 1-2   A Cloud Applications and Services IoT System

A generic cloud applications and services IoT system can also be generally represented by the following framework as shown in Figure 1-3.

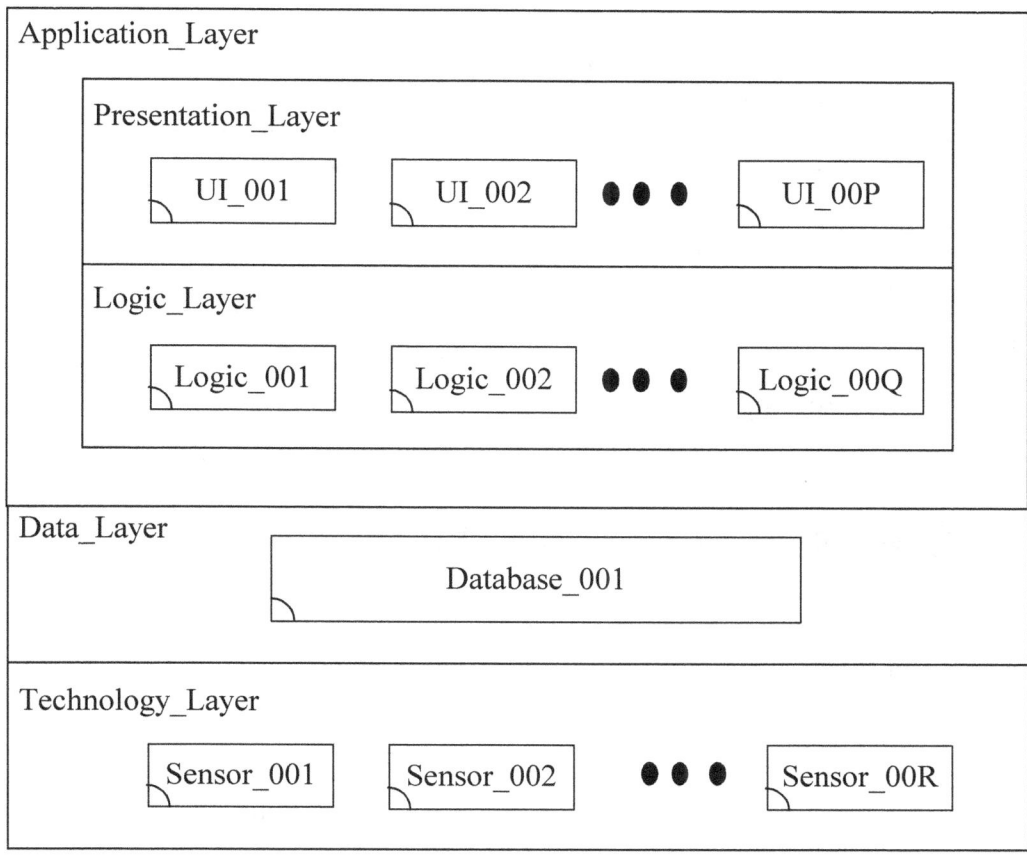

Figure 1-3   Generic Cloud Applications and Services IoT System

In the above figure, *Presentation_Layer* and *Logic_Layer* are sub-layers of *Application_Layer*. *Presentation_Layer* which describes the visual appearance of user interfaces contains many user interface (UI) components; *Logic_Layer* which describes the business logic or domain logic that encodes the real-world business rules contains many business logic components; *Data_Layer* which describes the system's logical and physical data assets contains a database component; *Technology_Layer* which describes the hardware and network infrastructure needed to support the deployment of data and applications contains many sensor components.

Behaviors of PSCASIS consist of: a) *Registering_Patient_Account* behavior, b) *Sensing_Patient_Vital_Signs* behavior, c) *Alerts_Notifying* behavior, d) *Recording_Emergency_Response_Starting_Time* behavior and e) *Recording_Emergency_Response_End_Time* behavior.

1-1 Behavior of Registering_Patient_Account

Each patient needs to register to PSCASIS to get the patient surveillance service, as shown in Figure 1-4.

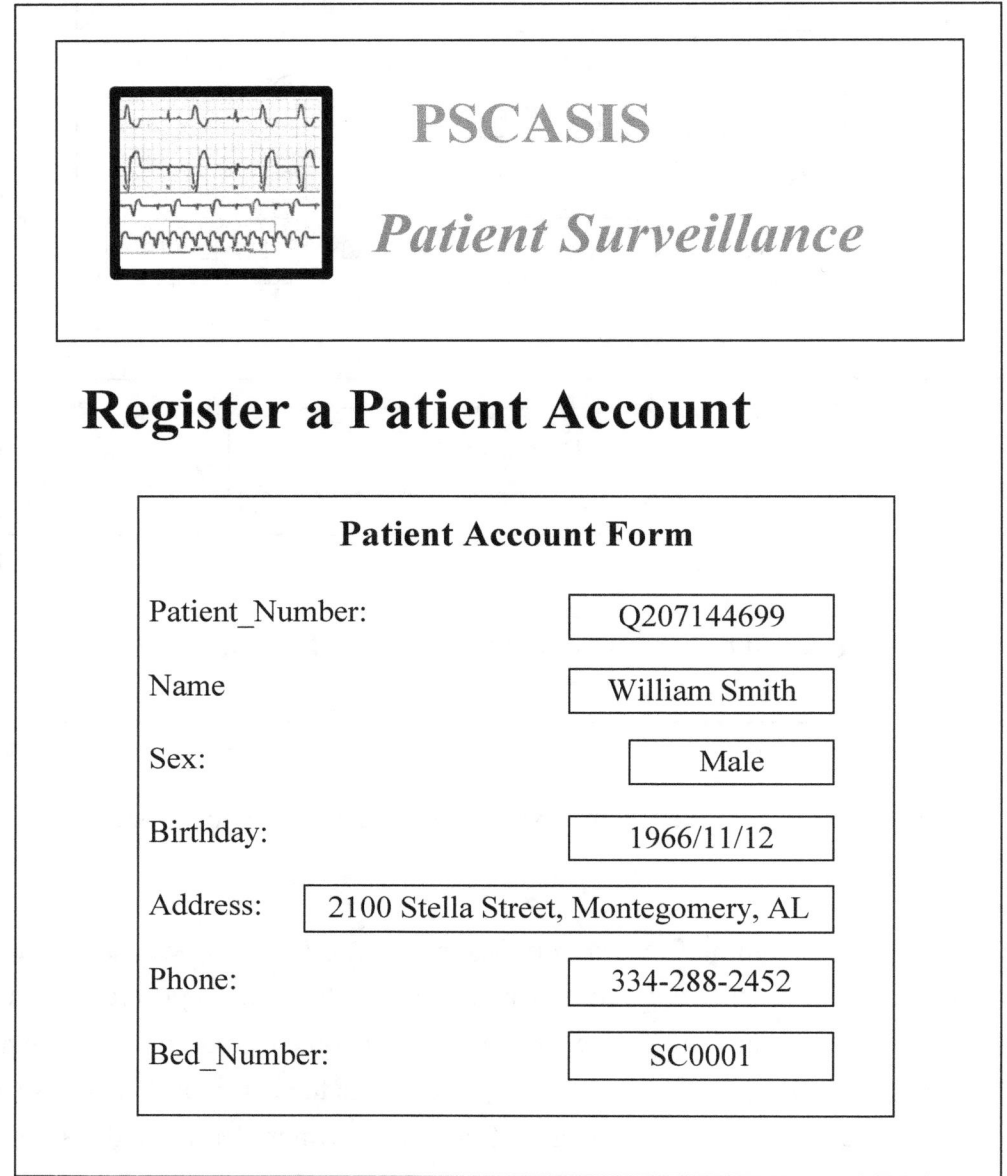

Figure 1-4   Registering a Patient Account to *PSCASIS*

In the behavior of *Registering_Patient_Account*, a nurse shall use the *Patient_Account_Registering_UI* component to input the corresponding data for this patient account registration. After that, the patient account registration data will be saved to the *PSCASIS_Database* component.

1-2 Behavior of Sensing_Patient_Vital_Signs

PSCASIS will sense and monitor every patient vital signs all the time, as shown in Figure 1-5..

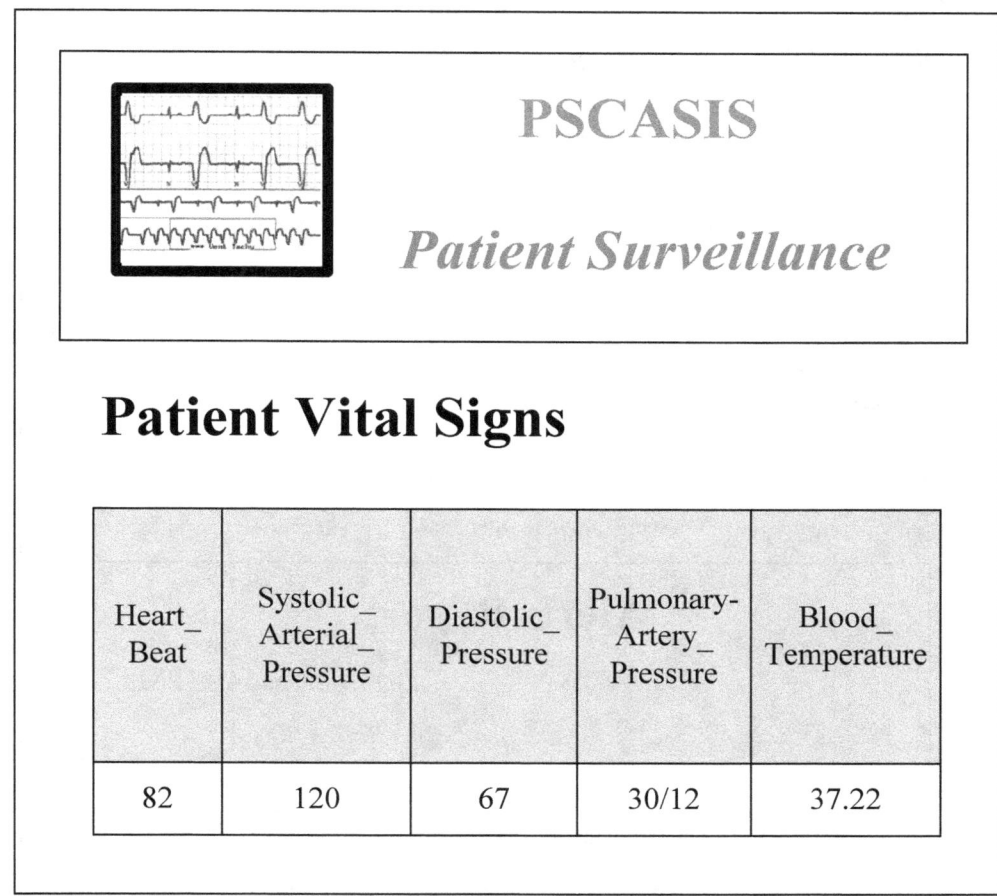

Figure 1-5   Vital Signs of Every Patient

Once the patient viral signs have been sensed, the sensor will pass the vital signs information to the *Patient_Vital_Signs_Daemon* component. After that, the *Patient_Vital_Signs_Daemon* component will communicate with the *PSCASIS_Database* component to record the patient vital signs information.

1-3 Behavior of Alerts_Notifying

By constantly (about 3 minutes) analyzing the vital signs of all patients, an alert will be triggered if there is any abnormal situation.

In the behavior of *Alerts_Notifying*, an alerts report is displayed on the *Alerts_Notifying_UI* component screen as shown in Figure 1-6.

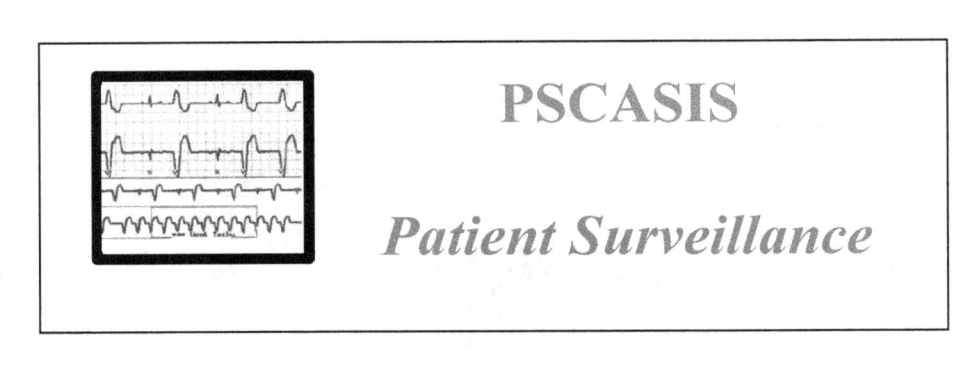

Patient_Number	Name	Alert_Code	Emergency_Response
A207144699	Grace Trump	10000	YES
B411277322	John Richardson	10000	NOT YET
B125670055	Tom Nixon	00010	NOT YET
C243112288	Lee Shelton	00011	YES

Figure 1-6　Displaying the Alerts

1-4 Behavior of Recording_Emergency_Response_Starting_Time

If there is any abnormal situation, alerts are displayed on the screen for a nurse to take an emergency response. Emergency response is the organizing, coordinating and directing of available resources in order to respond to the alert and bring the emergency under control.

Once the emergency response has been started, a nurse should record it on PSCASIS as shown in Figure 1-7.

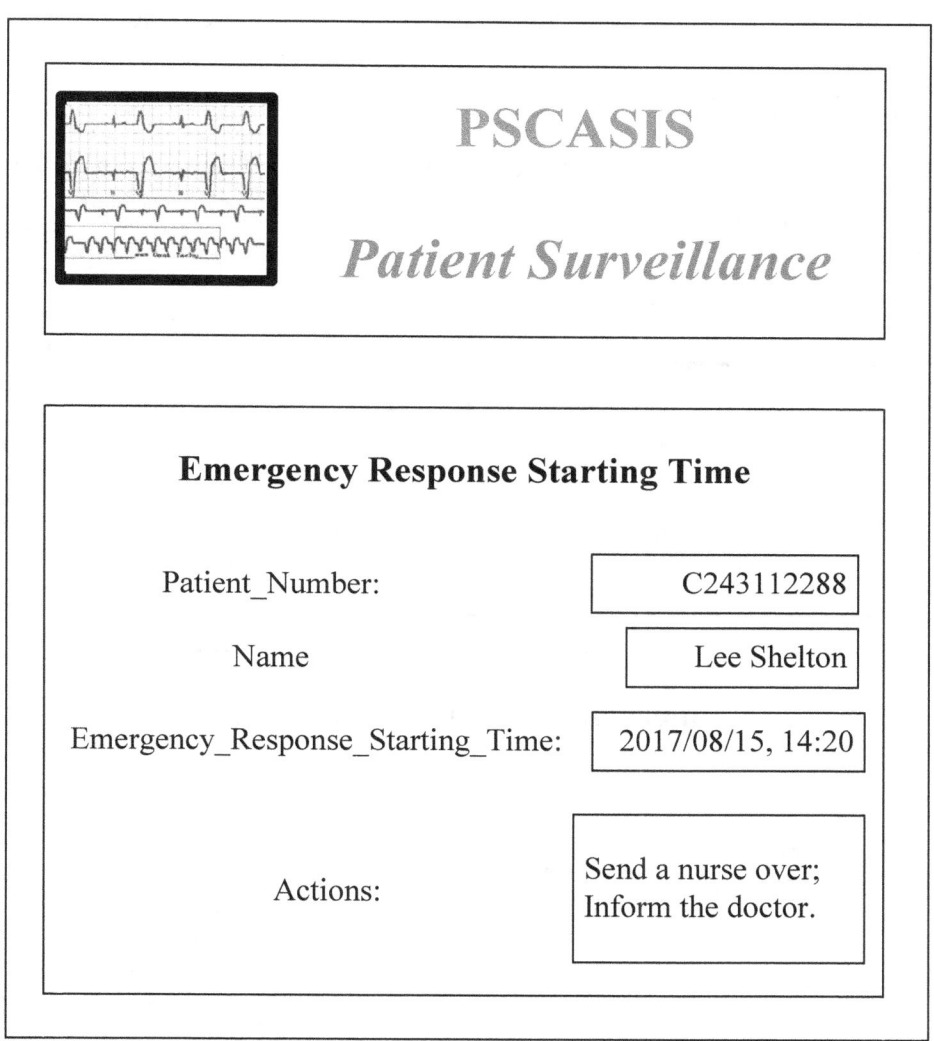

Figure 1-7 Recording the Emergency Responses Starting Time

In the behavior of *Recording_Emergency_Response_Starting_Time*, a nurse shall use the *Emergency_Response_Starting_Time_UI* component to input the corresponding data for this emergency response starting time. After that, the emergency response starting time data will be saved to the *PSCASIS_Database* component.

1-5 Behavior of Recording_Emergency_Response_End_Time

If there is any abnormal situation, alerts are displayed on the screen for a nurse to take an emergency response. Emergency response is the organizing, coordinating and directing of available resources in order to respond to the alert and bring the emergency under control.

Once the emergency response has been accomplished, a nurse should record it on PSCASIS as shown in Figure 1-8.

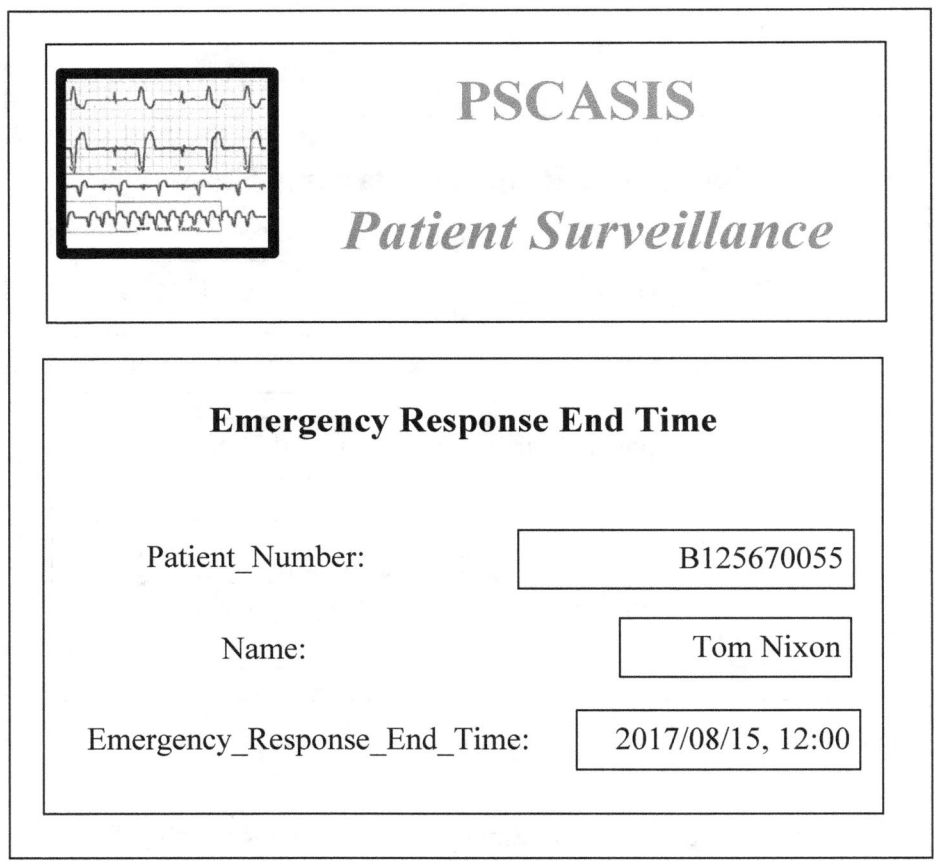

Figure 1-8 Recording the Emergency Responses End Time

In the behavior of *Recording_Emergency_Response_End_Time*, a nurse shall use the *Emergency_Response_End_Time_UI* component to input the corresponding data for this emergency response end time. After that, the emergency response end time data will be saved to the *PSCASIS_Database* component.

Chapter 2: Introduction to General Systems Theory 2.0

A system comprises multiple views such as strategy/version n, strategy/version n+1, concept, analysis, design, implementation, structure, behavior and input/output data views. A systems model is required to describe and represent all these multiple views.

The systems model describes and represents the system multiple views possibly using two different approaches. The first one is the non-architectural approach and the second one is the architectural approach. The non-architectural approach respectively picks a model for each view. The architectural approach, instead of picking many different models, will use only one single multiple views coalescence (MVC) model.

In general, MVC architecture is synonymous with the systems architecture. Since structure and behavior views are the two most prominent ones among multiple views, integrating the structure and behavior views becomes a superb approach for integrating multiple views of a system. In other words, structure-behavior coalescence (SBC) leads to the coalescence of multiple views. Therefore, we conclude that SBC architecture is also synonymous with the systems architecture.

2-1 Multiple Views of a System

In general, a system is extremely complex that it consists of several evolution&motivation views such as strategy/version n and strategy/version n+1 views; it also consists of various multi-level (hierarchical) views such as concept, analysis, design and implementation views; it also consists of many systemic views such as structure, behavior and input/output data views [Kend10, Pres09, Somm06].

Figure 2-1 shows that in a system all these strategy/version n, strategy/version n+1, concept, analysis, design, implementation, structure, behavior and input/output data views represent the multiple views of a system.

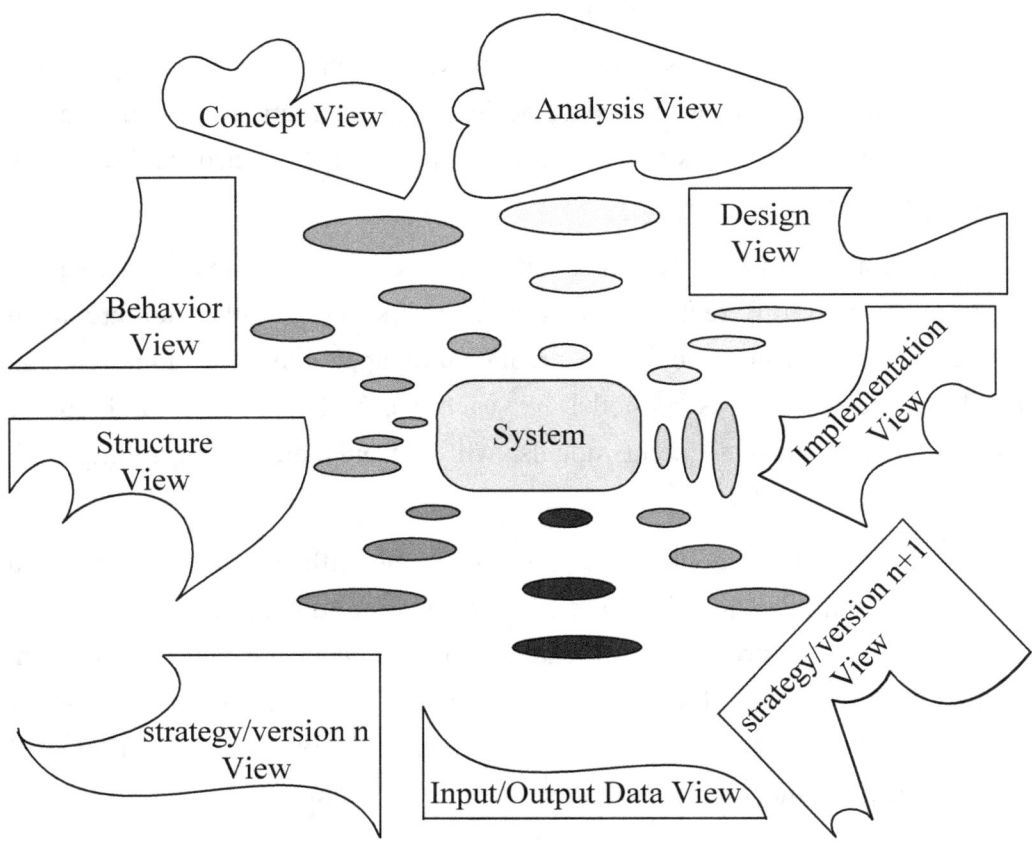

Figure 2-1   Multiple Views of a System

Among the above multiple views, the structure and behavior views are perceived as the two prominent ones. The structure view focuses on the systems structure which is described by components and their composition while the behavior view concentrates on the systems behavior which involves interactions (or handshakes) among the external environment's actors and components. Strategy/version n, strategy/version n+1, concept, analysis, design, implementation and input/output data views are considered to be other views as shown in Figure 2-2.

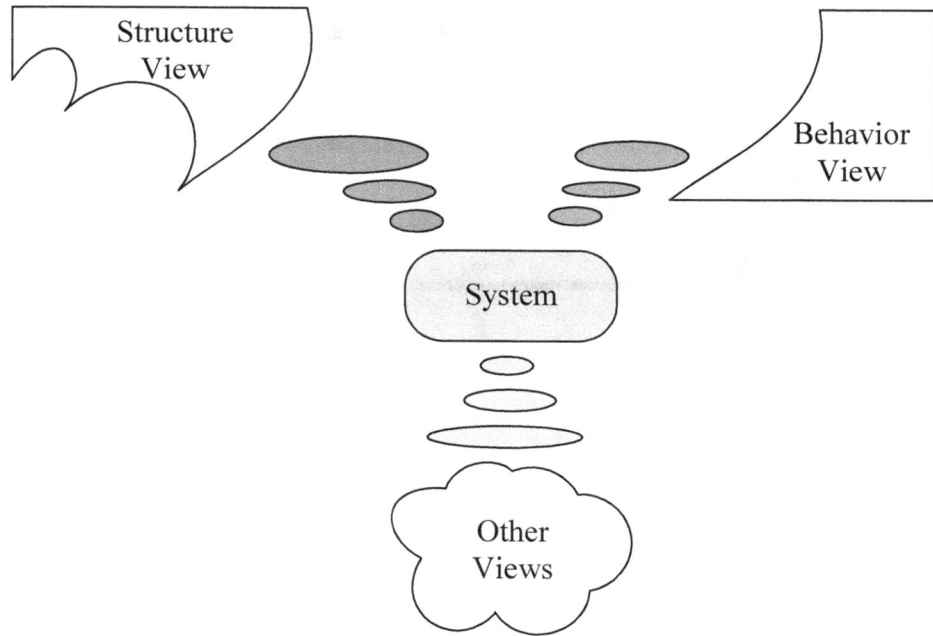

Figure 2-2   Structure, Behavior and Other Views

Accordingly, a system is defined in Figure 2-3 as an integrated whole of that system's multiple views, i.e., structure, behavior and other views, embodied in its assembled components, their interactions (or handshakes) with each other and the environment. Components are sometimes named as non-aggregated systems, parts, entities, objects and building blocks [Chao14a, Chao14b, Chao14c, Chec99].

> A system, hopefully is an integrated whole of that system's multiple views, i.e., structure, behavior and other views, embodied in its assembled components, their interrelationships with each other and the environment.

Figure 2-3   Definition of a System

Since multiple views are embodied in a system's assembled components which belong to the systems structure, they shall not exist alone. Multiple views must be loaded on the systems structure just like a cargo is loaded on a ship as shown in Figure 2-4. There will be no multiple views if there is no systems structure. Stand-alone multiple views are not meaningful.

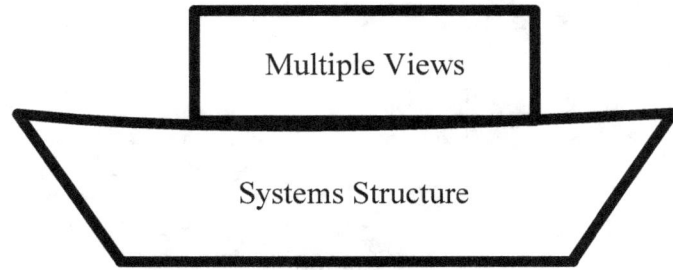

Figure 2-4   Multiple Views Must be Loaded on the Systems Structure

2-2 Non-Architectural Approach versus Architectural Approach

A system is exceptionally complex that it includes multiple views such as strategy/version n, strategy/version n+1, concept, analysis, design, implementation, structure, behavior and input/output data views.

The systems model describes and represents the system multiple views possibly using two different approaches. The first one is the non-architectural approach and the second one is the architectural approach.

The non-architectural approach, also known as the model multiplicity approach [Dori95, Dori02, Dori16], respectively picks a model for each view as shown in Figure 2-5, the strategy/version n view has the strategy/version n model, the strategy/version n+1 view has the strategy/version n+1 model, the concept view has the concept model, the analysis view has the analysis model, the design view has the design model, the implementation view has the implementation model, the structure view has the structure model, the behavior view has the behavior model, and the input/output data view has the input/output data model. These multiple models are separated, always inconsistent with each other, and then become the primary cause of model multiplicity problems [Dori95, Dori02, Dori16, Pele02, Sode03].

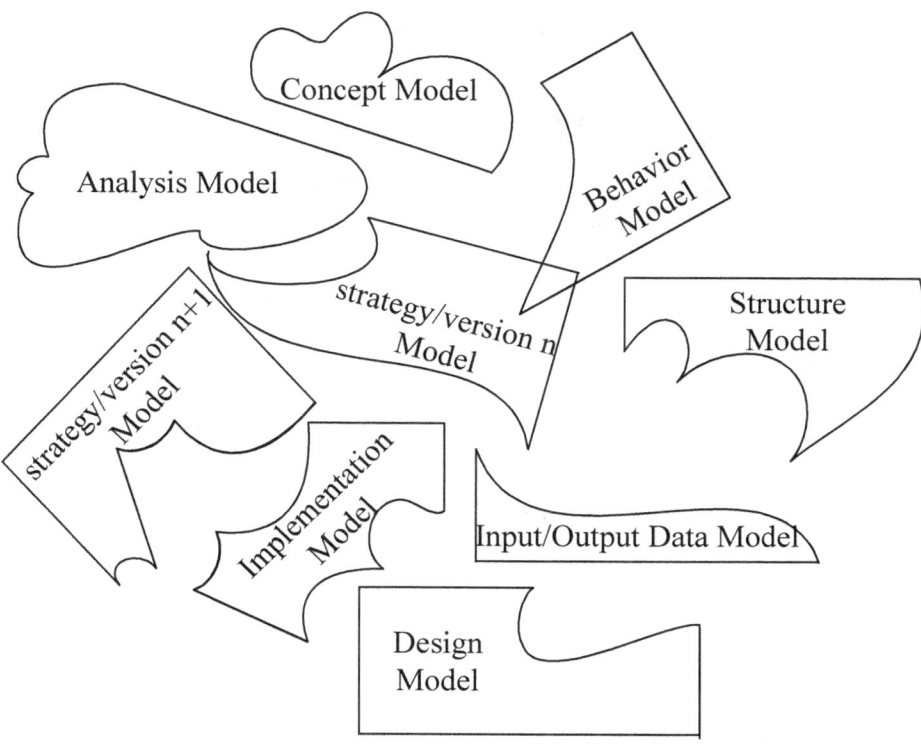

Figure 2-5 The Non-architectural Approach Picks a Model for Each View

The architectural approach, also known as the model singularity approach [Dori95, Dori02, Dori16, Pele02, Sode03], instead of picking many heterogeneous and separated models, will use only one single coalescence model as shown in Figure 2-6. The strategy/version n, strategy/version n+1, concept, analysis, design, implementation, structure, behavior and input/output data views are all integrated in this multiple views coalescence (MVC) model of systems architecture (SA) [Chao14a, Chao14b, Chao14c, Chao15a, Chao15b, Chao16, Chao17a, Chao17b, Chao17c, Chao17d, Chao17e, Chao17f].

Figure 2-6   Systems Architecture Uses a Coalescence Model

Figure 2-5 has many models. Figure 2-6 has only one model. Comparing Figure 2-5 with Figure 2-6, we unquestionably conclude that an integrated, holistic, united, coordinated, coherent and coalescence model is more favorable than a collection of many heterogeneous and separated models.

2-3 Definition of Systems Architecture

Involved systems are extremely complex in every aspect so that each stakeholder needs a blueprint or model to capture their essential structures and behaviors. Systems architecture is such a blueprint or model.

There are several well-know definitions of systems architecture [Burd10, Craw15, Dam06, Maie09, O'Rou03, Roza11]. ANSI/IEEE 1471-2000 defines systems architecture as: "the fundamental organization of a system, embodied in its components, their relationships to each other and the environment, and the principles governing its design and evolution." The Open Group defines systems architecture as either "a formal description of a system, or a detailed plan of the system at component level to guide its implementation," or as "the structure of components, their interrelationships, and the principles and guidelines governing their design and evolution over time" [Rayn09, Toga08].

Concluding the above definitions, we now give systems architecture a definition of our own as shown in Figure 2-7.

> Systems architecture is an integrated whole of a system's multiple views, i.e., structure, behavior and other views, embodied in its assembled components, their interactions with each other and the environment, and the principles and guidelines governing its design and evolution.

Figure 2-7   Definition of Systems Architecture

From the above definition, we find out that systems architecture is an integrated whole of a system's multiple views, i.e., structure, behavior and other views, embodied in its assembled components, their interactions (or handshakes) with each other and the environment, and the principles and guidelines governing its design and evolution. That is, systems architecture is an integrated and coalescence model of multiple views. In this coalescence model, structure, behavior and other views are all included in it as shown in Figure 2-8. We do not supply each view a respective model in this systems architecture coalescence model.

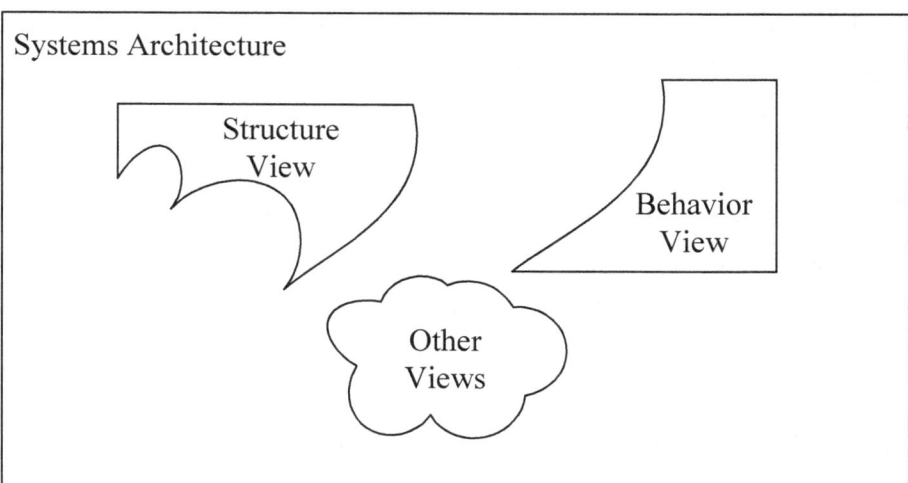

Figure 2-8   All Multiple Views are Included in This Systems Architecture

Since multiple views are embodied in a system's assembled components which belong to the structure view, they shall not exist alone. Multiple views must be loaded on the structure view just like a cargo is loaded on a ship as shown in Figure 2-9. There will be no multiple views if there is no structure view. Stand-alone multiple views are not meaningful.

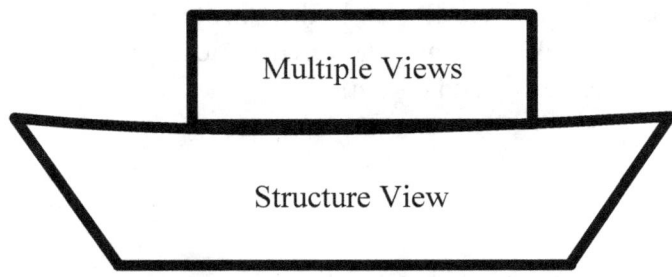

Figure 2-9   Multiple Views Must be Loaded on the Structure View

2-4 Architecture Description Language

An architecture description is a formal description and representation of a system. A description of the systems architecture has to grasp the essence of the system and its details at the same time. In other words, an architecture description not only provides an overall picture that summarizes the whole system, but also contains enough detail that the system can be constructed and validated.

The language for architecture description is called the architecture description language (ADL) [Bass03, Clem02, Clem10, Dike01, Roza11, Shaw96, Tayl09]. An ADL is a special kind of language used in describing the architecture of a system.

Since the architectural approach uses a coalescence model for all multiple views of a system, the foremost duty of ADL is to make the strategy/version n, strategy/version n+1, concept, analysis, design, implementation, structure, behavior and input/output data views all integrated and coalesced within this architecture description.

2-5 Multiple Views Coalescence to Achieve the Systems Architecture

Systems architecture has been defined as a coalescence model of multiple views. Multiple views coalescence (MVC) uses only a single coalescence model as shown in Figure 2-10. Strategy/version n, strategy/version n+1, concept, analysis, design, implementation, structure, behavior and input/output data views are all integrated in this MVC architecture.

Figure 2-10   MVC Architecture

Generally, MVC architecture is synonymous with the systems architecture. In other words, multiple views coalescence sets a path to achieve the systems architecture as shown in Figure 2-11.

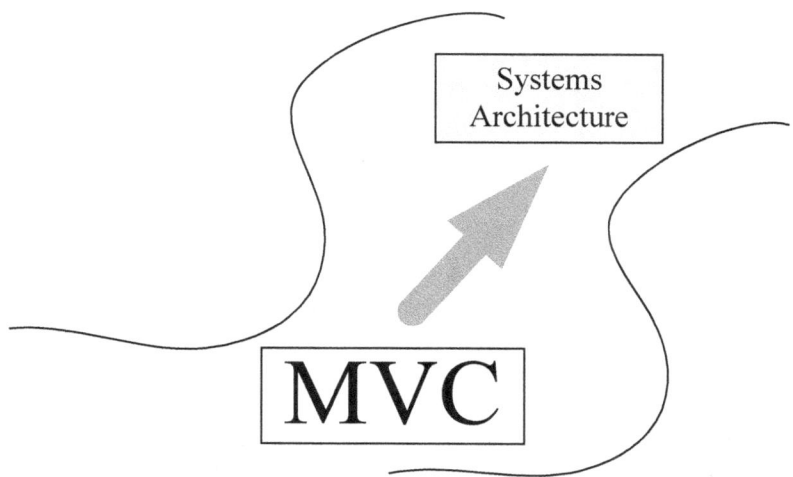

Figure 2-11   MVC to Achieve the Systems Architecture

In the MVC architecture, multiple views must be attached to or built on the systems structure. In other words, multiple views shall not exist alone; they must be loaded on the systems structure just like a cargo is loaded on a ship as shown in Figure 2-12. There will be no multiple views if there is no systems structure. Stand-alone multiple views are not meaningful.

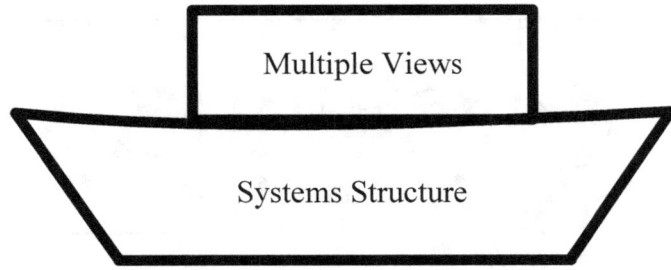

Figure 2-12   Multiple Views Must be Loaded on the Systems Structure

2-6 Integrating the Systems Structures and Systems Behaviors

By integrating the systems structure and systems behavior, we obtain structure-behavior coalescence (SBC) within the system as shown in Figure 2-13.

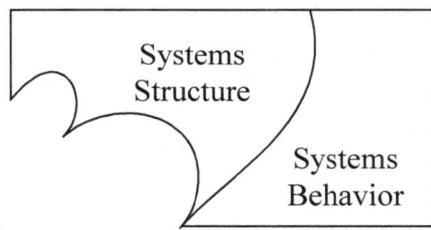

Figure 2-13   Structure-Behavior Coalescence

Structure-behavior coalescence has never been used in any systems model (SM) for systems development except the SBC architecture and object-process methodology (OPM) [Dori95, Dori02, Dori16, Pele02, Sode03]. There are many advantages to use the structure-behavior coalescence approach to integrate the systems structure and systems behavior.

SBC architecture uses a single coalescence model as shown in Figure 2-14. Systems structures and systems behaviors are integrated in this SBC architecture.

Figure 2-14   SBC Architecture

Since systems structures and systems behaviors are so tightly integrated, we sometimes claim that the core theme of SBC architecture is: Systems Architecture = Systems Structure + Systems Behavior, as shown in Figure 2-15.

Figure 2-15   Core Theme of SBC Architecture

So far, systems behaviors are separated from systems structures in most cases [Pres09, Somm06]. For example, the well-known structured systems analysis and design (SSA&D) approach uses structure charts (SC) to represent the systems structure and data flow diagrams (DFD) to represent the systems behavior [Denn08, Kend10, Your99]. SC and DFD are two different models. They are so separated like that there is the "Atlantic Ocean" between them, as shown in Figure 2-16.

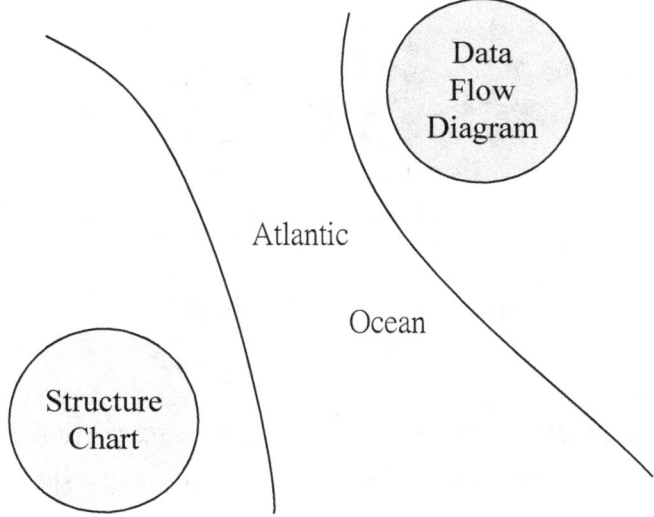

Figure 2-16   Two Heterogeneous and Separated Models

2-7 Structure-Behavior Coalescence to Facilitate Multiple Views Coalescence

Since structure and behavior views are the two most prominent ones among multiple views, integrating the structure and behavior views is clearly the best way to integrate multiple views of a system. In other words, structure-behavior coalescence facilitates multiple views coalescence as shown in Figure 2-17.

Figure 2-17   SBC Facilitates MVC

2-8 Structure-Behavior Coalescence to Achieve the Systems Architecture

Figure 2-11 declares that multiple views coalescence sets a path to achieve the desired systems architecture with the most efficient approach. Figure 2-17 declares that structure-behavior coalescence facilitates multiple views coalescence.

Combining the above two declarations, we conclude that structure-behavior coalescence sets a path to achieve the systems architecture as shown in Figure 2-18. In this case, SBC architecture is also synonymous with the systems architecture.

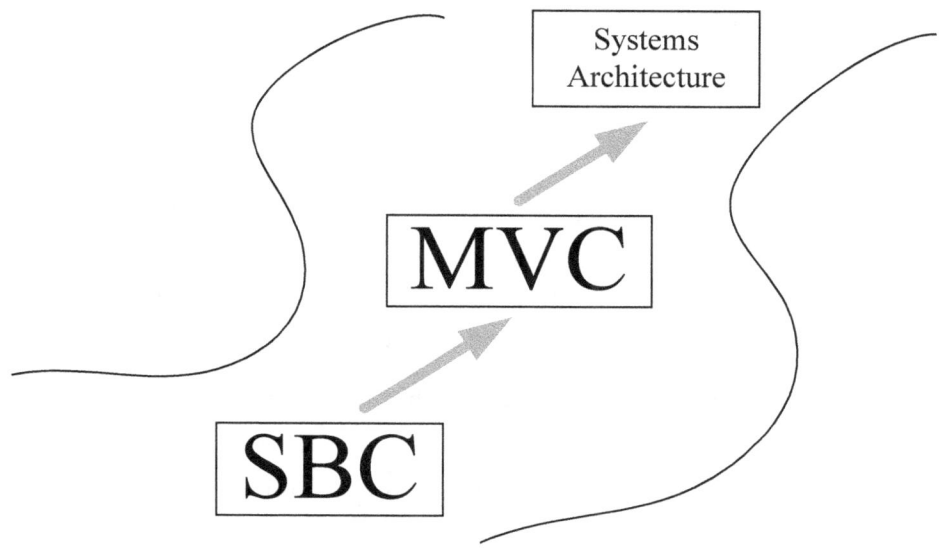

Figure 2-18   SBC to Achieve the Systems Architecture

SBC architecture strongly demands that the structure and behavior views must be coalesced and integrated. This never happens in other architectural approaches such as Zachman Framework [O'Rou03], The Open Group Architecture Framework (TOGAF) [Rayn09, Toga08], Department of Defense Architecture Framework (DoDAF) [Dam06] and Unified Modeling Language (UML) [Rumb91]. Zachman Framework does not offer any mechanism to integrate the structure and behavior views. TOGAF, DoDAF and UML do not, either.

In the SBC architecture, the systems behavior must be attached to or built on the systems structure. In other words, the systems behavior can not exist alone; it must be loaded on the systems structure just like a cargo is loaded on a ship as shown in Figure 2-19. There will be no systems behavior if there is no systems structure. A stand-alone systems behavior is not meaningful.

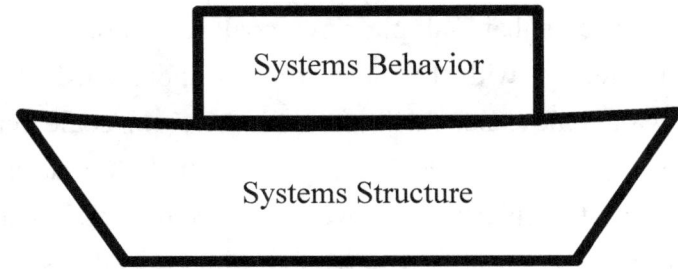

Figure 2-19   Systems Behavior is Loaded on the Systems Structure

2-9 Using SBC-ADL to Construct the Systems Architecture

An architecture description language (ADL) is a special kind of language used in describing the architecture of a system [Shaw96, Tayl09].

A description of the systems architecture has to grasp the essence of a system and its details at the same time. In other words, a systems architecture description not only provides an overall picture that summarizes the system, but also contains enough detail that the system can be constructed and validated.

SBC-ADL uses six fundamental diagrams to describe the integration of systems structure and systems behavior of a system. These diagrams, as shown in Figure 2-20, are: a) architecture hierarchy diagram (AHD), b) framework diagram (FD), c) component channel diagram (CChD), d) component connection diagram (CCoD), e) structure-behavior coalescence diagram (SBCD) and f) interaction flow diagram (IFD).

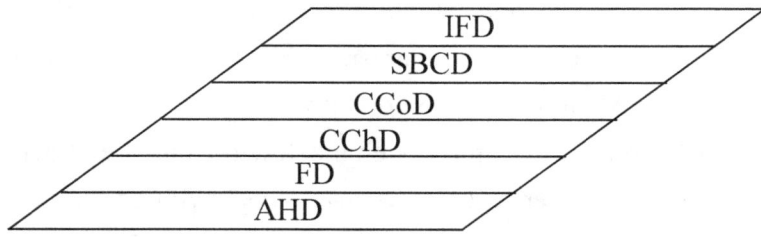

Figure 2-20   Six Fundamental Diagrams of SBC-ADL

SBC-ADL uses AHD, FD, CChD, CCoD, SBCD and IFD to depict the systems structure and systems behavior of a system as shown in Figure 2-21.

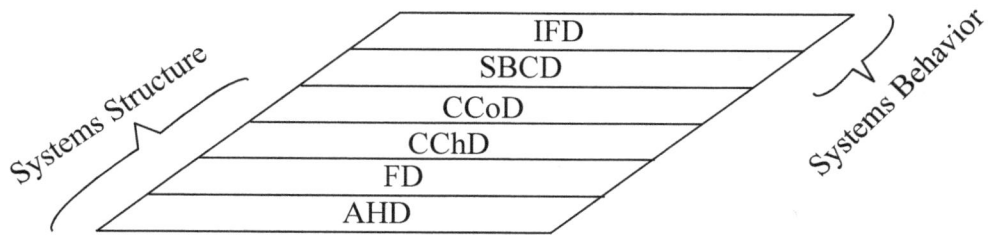

Figure 2-21   Systems Structure and Systems Behavior of a System

Examining the SBC-ADL approach, we find out that it depicts the systems structure first and then depicts the systems behavior later, not the other way around. The reason SBC-ADL does so lies in that the systems behavior must be attached to or built on the systems structure. With the systems structure and attached systems behavior, then, we can smoothly get the systems architecture as shown in Figure 2-22.

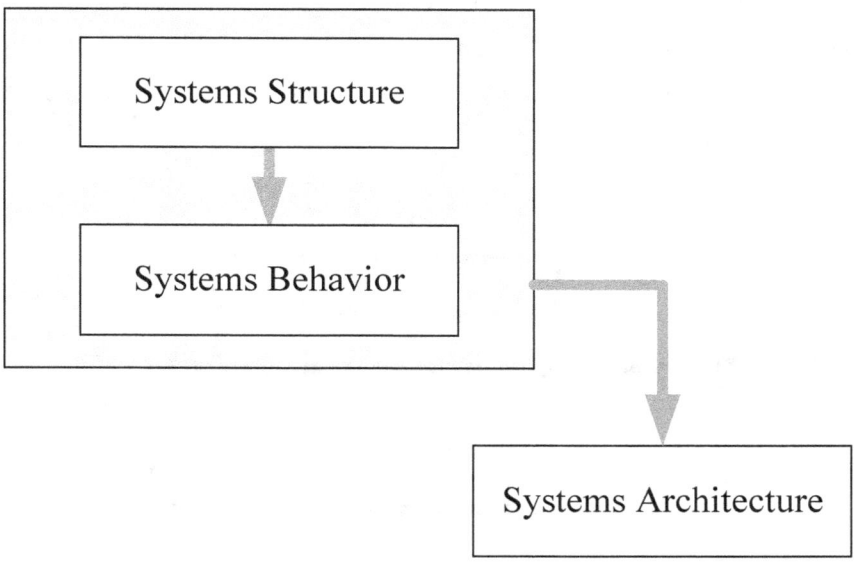

Figure 2-22   Systems Behavior is Attached to the Systems Structure

Let us ask the opposite question. Can the systems structure be attached to or built on the systems behavior? The answer is "No" as shown in Figure 2-23.

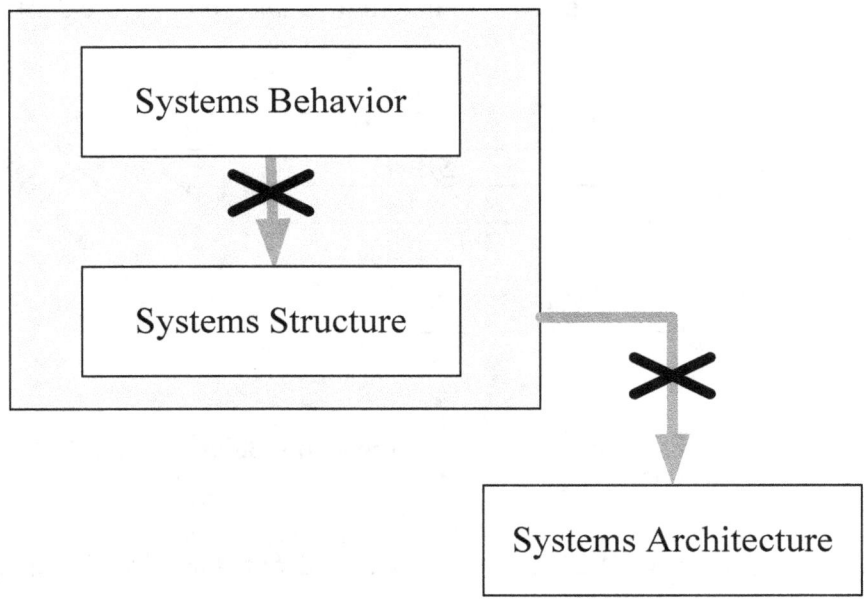

Figure 2-23 Systems Structure is not Attached to the Systems Behavior

In the SBC-ADL, systems behavior must be attached to or built on the systems structure. In other words, the systems behavior shall not exist alone; it must be loaded on the systems structure just like a cargo is loaded on a ship as shown in Figure 2-24. There will be no systems behavior if there is no systems structure. A stand-alone systems behavior is not meaningful.

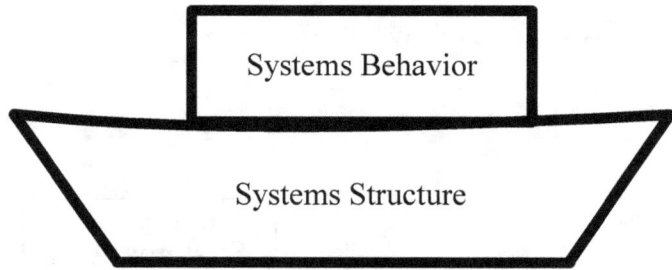

Figure 2-24 Systems Behavior Must be Loaded on the Systems Structure

AHD, FD, CChD and CCoD belong to systems structure. SBCD and IFD belong to systems behavior. Concluding the above discussion, we perceive that SBC-ADL will describe AHD, FD, CChD and CCoD first then describe SBCD and IFD later when it constructs the systems architecture of a system.

2-10 SBC Model Singularity

Channel-Based Single-Queue SBC Process Algebra (C-S-SBC-PA) [Chao17a], Channel-Based Multi-Queue SBC Process Algebra (C-M-SBC-PA) [Chao17b], Channel-Based Infinite-Queue SBC Process Algebra (C-I-SBC-PA) [Chao17c], Operation-Based Single-Queue SBC Process Algebra (O-S-SBC-PA) [Chao17d], Operation-Based Multi-Queue SBC Process Algebra (O-M-SBC-PA) [Chao17e] and Operation-Based Infinite-Queue SBC Process Algebra (O-I-SBC-PA) [Chao17f] are the six specialized SBC process algebras. The SBC process algebra (SBC-PA) shown in Figure 2-25 is a model singularity approach.

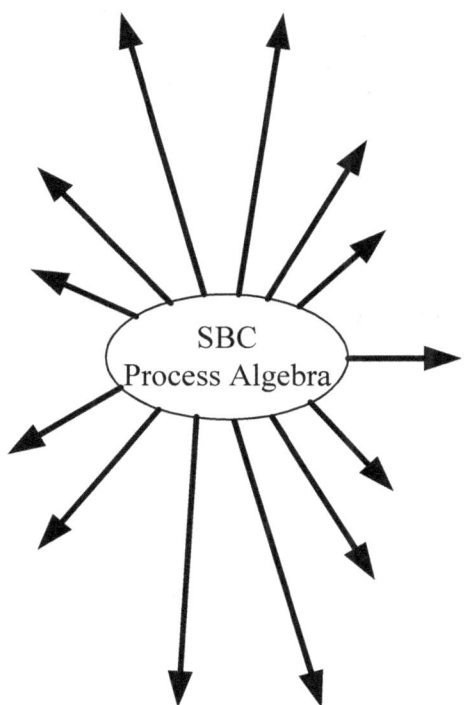

Figure 2-25   SBC-PA is a Model Singularity Approach.

The SBC architecture description language (SBC-ADL) is also a model singularity approach. With SBC mind set sitting in the kernel, the SBC-ADL single model shown in Figure 2-26 is therefore able to represent all structural views such as architecture hierarchy diagram (AHD), framework diagram (FD), component channel diagram (CChD), component connection diagram (CCoD) and behavioral views such as structure-behavior coalescence diagram (SBCD), interaction flow diagram (IFD).

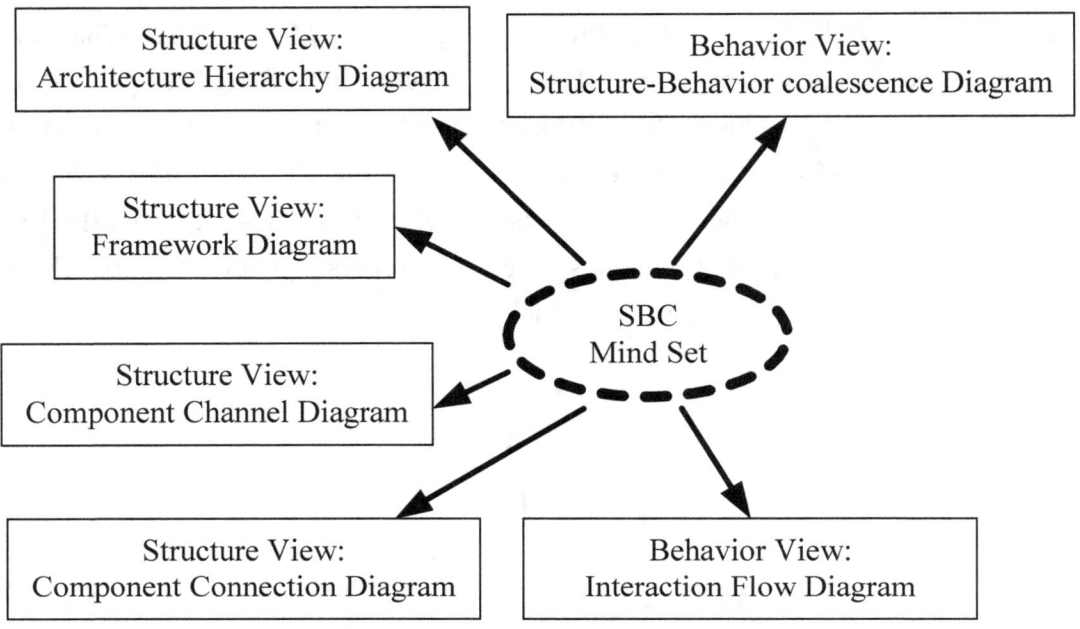

Figure 2-26 SBC-ADL is a Model Singularity Approach.

The combination of SBC process algebra (SBC-PA) and SBC architecture description language (SBC-ADL) is shown in Figure 2-27, again as a model singularity approach.

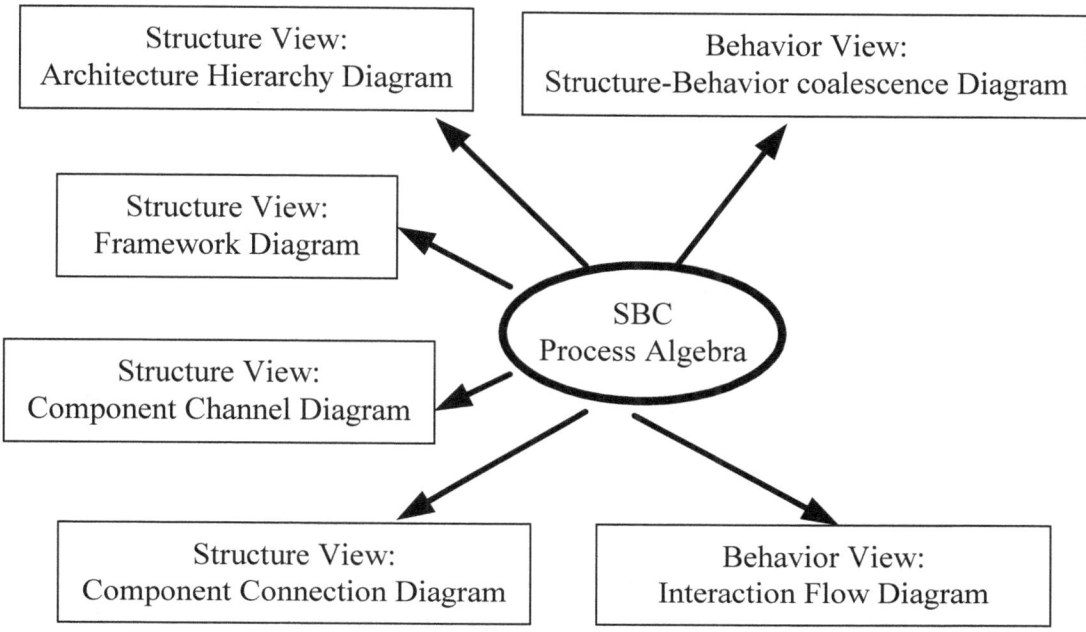

Figure 2-27   SBC Model is a Model Singularity Approach.

PART II: SBC ARCHITECTURE DESCRIPTION LANGUAGE

Chapter 3: Systems Structure

SBC-ADL uses the architecture hierarchy diagram, framework diagram, component channel diagram and component connection diagram to depict the systems structure of a system.

3-1 Architecture Hierarchy Diagram

Systems architects use an architecture hierarchy diagram (AHD) to define the multi-level (hierarchical) decomposition and composition of a system. AHD is the first fundamental diagram to achieve structure-behavior coalescence.

3-1-1 Decomposition and Composition

The following is an example of systems decomposition and composition. The *Computer* system consists of *Monitor*, *Keyboard*, *Mouse* and *Case*, as shown in Figure 3-1. The *Monitor*, *Keyboard*, *Mouse* and *Case* are subsystems comprising the *Computer* system.

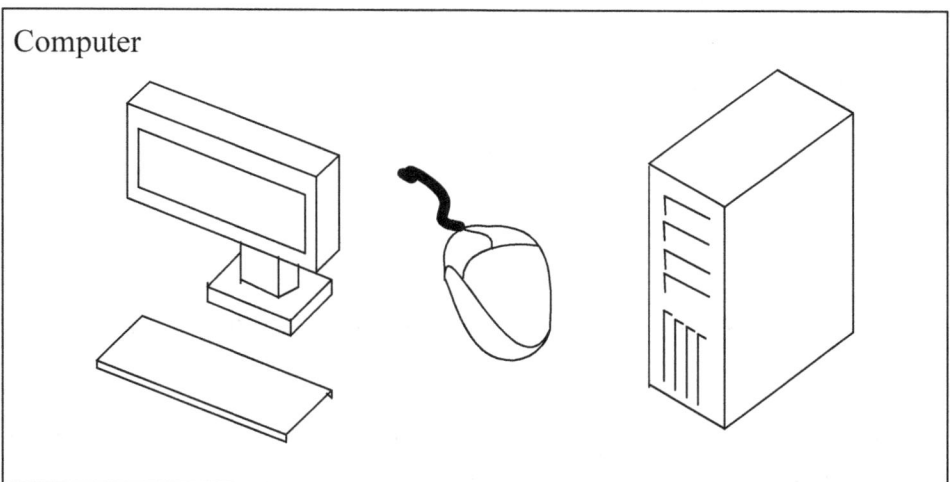

Figure 3-1   Decomposition and Composition of the *Computer* System

Another example indicates that the *Tree* system is composed of *Root* and *Stem*, as shown in Figure 3-2. In this example, we would say that the *Root* and *Stem* are subsystems, respectively, while the *Tree* system consists of its subsystems.

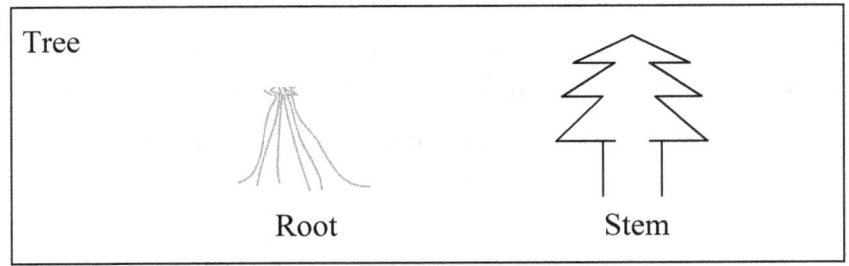

Figure 3-2   Decomposition and Composition of the *Tree* System

The last example demonstrates that the *SBC_Book* system is composed of *Chapter_1*, *Part_1* and *Part_2*, as shown in Figure 3-3. In this example, we would say that *Chapter_1*, *Part_1* and *Part_2* are subsystems, respectively while the *SBC_Book* system consists of its subsystems.

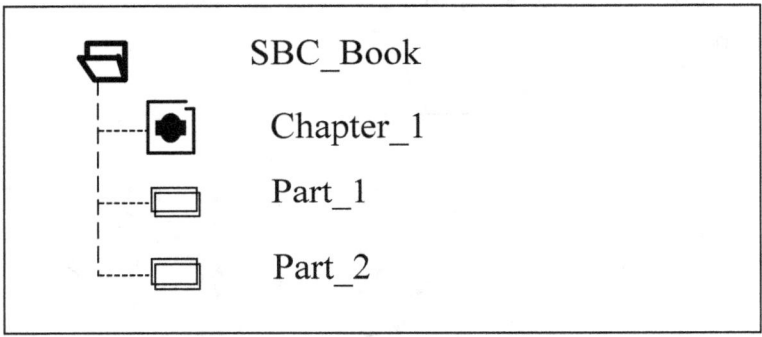

Figure 3-3   Decomposition and Composition of the *SBC_Book* System

Architecture hierarchy diagram (AHD) is used to define the decomposition and composition of a system. As an example, Figure 3-4 shows an AHD of the *Computer* system. We clearly observe that the *Computer* system is composed of *Monitor*, *Keyboard*, *Mouse* and *Case*.

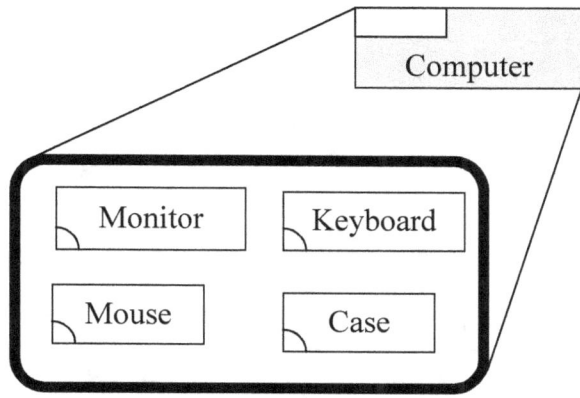

Figure 3-4   AHD of the *Computer* System

As a second example, Figure 3-5 shows an AHD of the *Tree* system. We clearly observe that the *Tree* system is composed of *Root* and *Stem*.

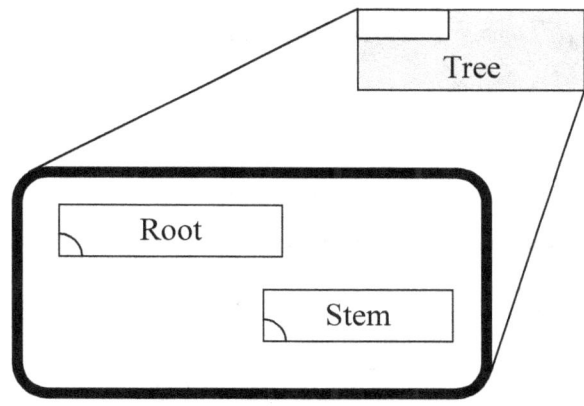

Figure 3-5   AHD of the *Tree* System

As a third example, Figure 3-6 shows an AHD of the *SBC_Book* system. We clearly observe that the *SBC_Book* is composed of *Chapter_1*, *Part_1* and *Part_2*.

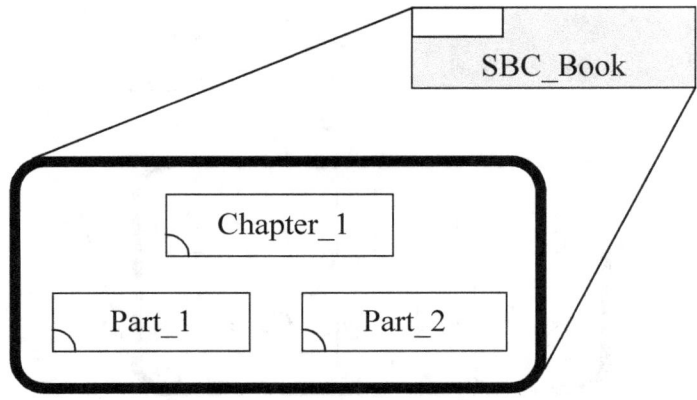

Figure 3-6   AHD of the *SBC_Book* system

3-1-2 Multi-Level Decomposition and Composition

The subsystem may also contain subsystems as we further decompose it. For example, *Case* is a subsystem of the *Computer*, and we can further decompose it into *Motherboard*, *Hard_Disk*, *Power_Supply* and *DVD_Disk*, as shown in Figure 3-7.

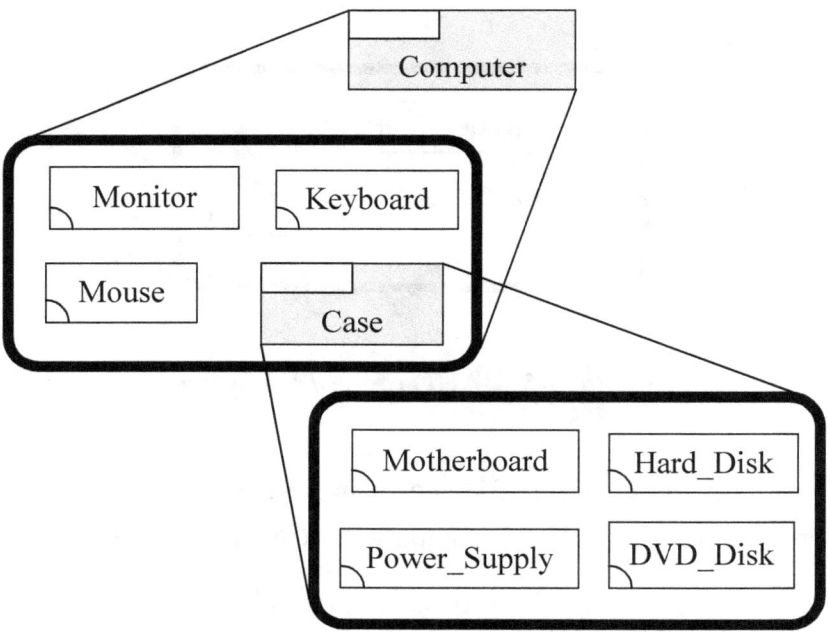

Figure 3-7 Multi-Level Decomposition/Composition of the *Computer* System

As a second example, *Stem* is a subsystem of the *Tree*, and we can further decompose it into *Trunk* and *Leaf*, as shown in Figure 3-8.

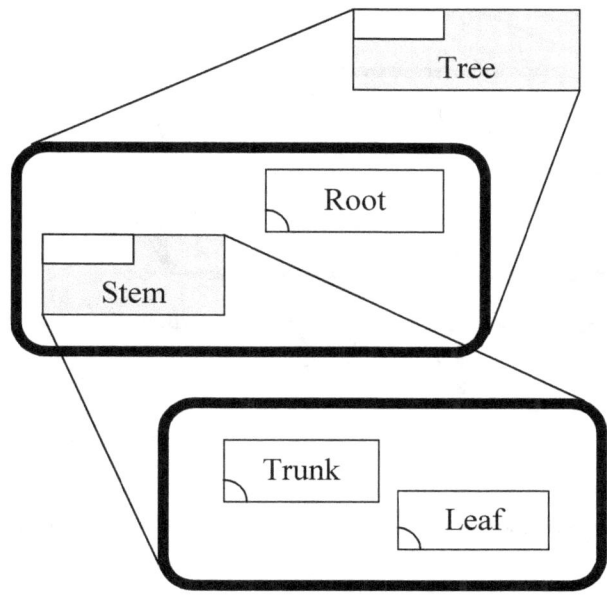

Figure 3-8   Multi-Level Decomposition/Composition of the *Tree* System

As a third example, *Part_1* is a subsystem of the *SBC_Book*, and we can further decompose it into *Chapter_2* and *Chapter_3*; *Part_2* is also a subsystem of the *SBC_Book*, and we can further decompose it into *Chapter_4* and *Chapter_5*, as shown in Figure 3-9.

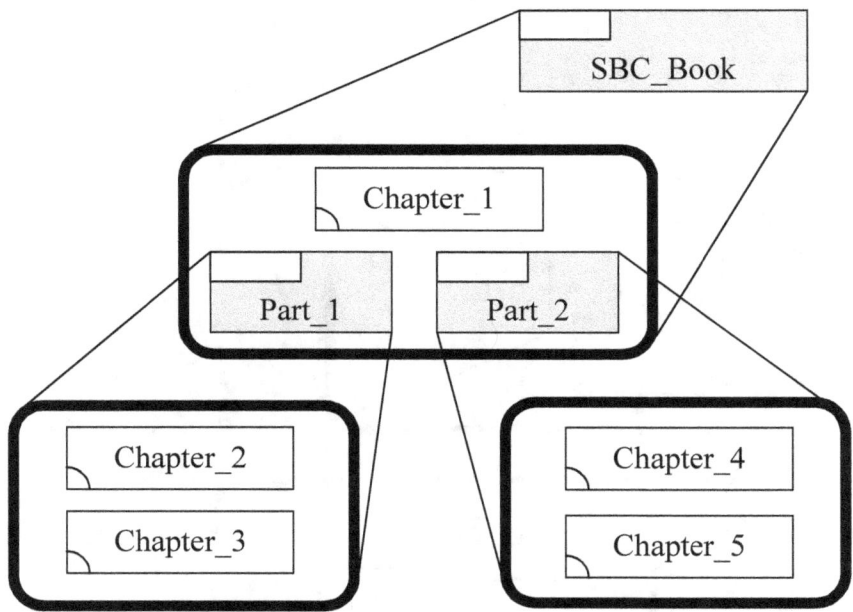

Figure 3-9 Multi-Level Decomposition/Composition of the *SBC_Book* System

Generally speaking, multi-level decomposition and composition of a system is applied often in constructing its architecture. To make a complex system look simple, the mechanism of multi-level composition and decomposition should always be used.

3-1-3 Aggregated and Non-Aggregated Systems

Any subsystem (at any level) involved with multi-level decomposition and composition of a system is either aggregated or non-aggregated. The definition of aggregated and non-aggregated systems is shown in Figure 3-10.

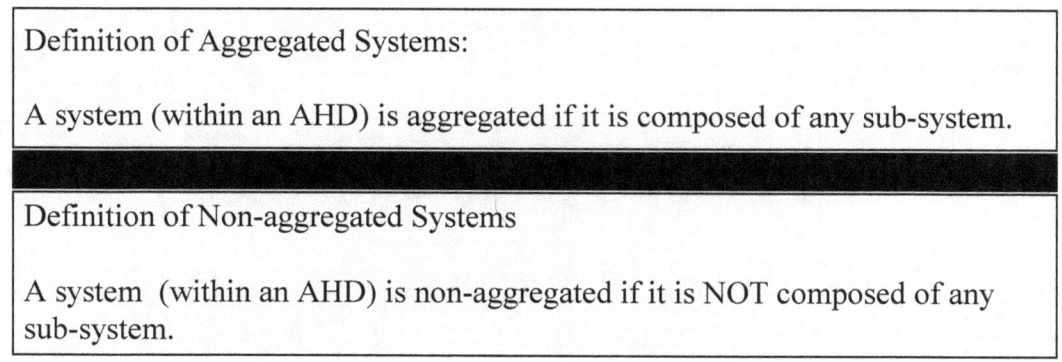

Figure 3-10   Definition of Aggregated and Non-aggregated Systems

Non-aggregated systems are sometimes referred to as components, parts, entities, objects and building blocks [Chao14a, Chao14b, Chao14c].

In the multi-level systems decomposition and composition, any system is either aggregated or non-aggregated, but not both. For example, in Figure 3-4, *Case* is a non-aggregated system, not an aggregated system. As an interesting contrast, in Figure 3-7, *Case* is an aggregated system, not a non-aggregated system.

As a second example, in Figure 3-5, *Stem* is a non-aggregated system, not an aggregated system. As an interesting contrast, in Figure 3-8, *Stem* is an aggregated system, not a non-aggregated system.

As a third example, in Figure 3-6, *Part_1 and Part_2* are non-aggregated systems, not aggregated systems. As an interesting contrast, in Figure 3-9, *Part_1* and *Part_2* are aggregated systems, not non-aggregated systems.

3-2 Framework Diagram

Framework diagram (FD) enables systems architects to examine the multi-layer (also referred to as multi-tier) decomposition and composition of a system. FD is the second fundamental diagram to achieve structure-behavior coalescence.

3-2-1 Multi-Layer Decomposition and Composition

Decomposition and composition of a system can also be represented in a multi-layer (or multi-tier) manner. We draw a framework diagram (FD) for the multi-layer decomposition and composition of a system.

As an example, Figure 3-11 shows a FD of the *Computer* system. In the figure, *Technology_SubLayer_2* contains *Monitor*, *Keyboard* and *Mouse*; *Technology_SubLayer_1* contains *Motherboard*, *Hard_Disk*, *Power_Supply* and *DVD_Disk*.

Figure 3-11 FD of the *Computer* System

As a second example, Figure 3-12 shows a FD of the *Tree* system. In the figure, *Technology_SubLayer_2* contains *Root*; *Technology_SubLayer_1* contains *Trunk* and *Leaf*.

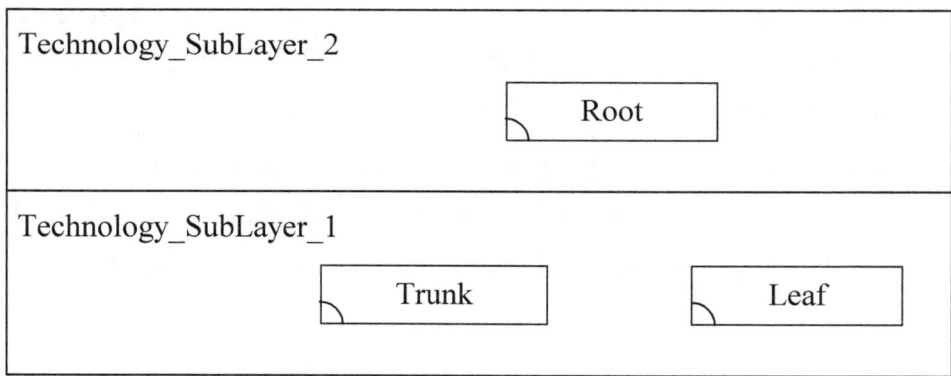

Figure 3-12 FD of the *Tree* System

As a third example, Figure 3-13 shows a FD of the *SBC_Book* system. In the figure, *Technology_SubLayer_2* contains *Chapter_1*; *Technology_SubLayer_1* contains *Chapter_2*, *Chapter_3*, *Chapter_4* and *Chapter_5*.

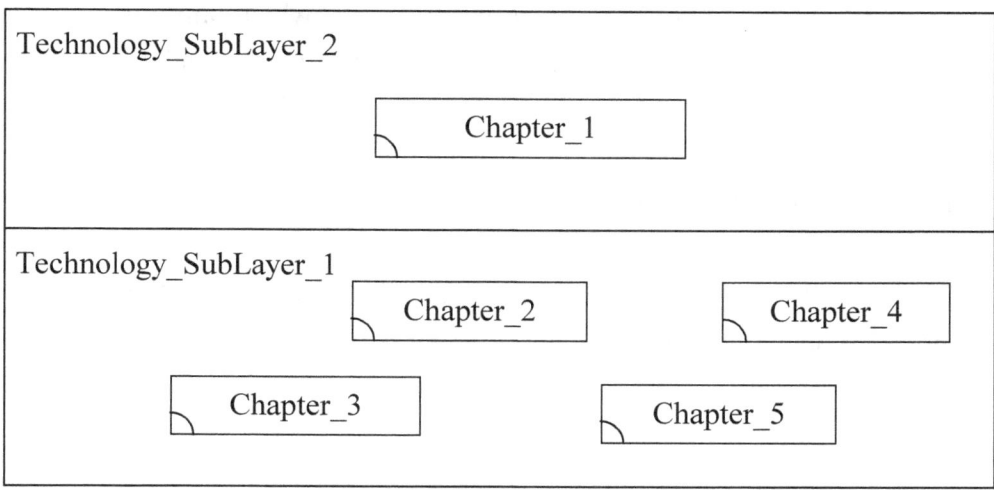

Figure 3-13 FD of the *SBC_Book* System

3-2-2 Only Non-Aggregated Systems Appearing in Framework Diagrams

Both aggregated and non-aggregated systems are displayed in the multi-level AHD decomposition and composition of a system. As an interesting contrast, only non-aggregated systems shall appear in the multi-layer FD decomposition and composition of a system.

For example, Figure 3-7 in the previous section shows an AHD of the *Computer* system in which both aggregated systems such as *Computer*, *Case* and non-aggregated systems such as *Monitor*, *Keyboard*, *Mouse*, *Motherboard*, *Hard_Disk*, *Power_Supply*, *DVD_Disk* are displayed. As an interesting contrast, Figure 3-11 in the previous section shows a FD of the *Computer* system in which only non-aggregated systems such as *Monitor*, *Keyboard*, *Mouse*, *Motherboard*, *Hard_Disk*, *Power_Supply* and *DVD_Disk* are displayed.

For a second example, Figure 3-8 in the previous section shows an AHD of the *Tree* system in which both aggregated systems such as *Tree*, *Stem* and non-aggregated systems such as *Root*, *Trunk*, *Leaf* are displayed. As an interesting contrast, Figure 3-12 in the previous section shows a FD of the *Tree* system in which only non-aggregated systems such as *Root*, *Trunk* and *Leaf* are displayed.

For a third example, Figure 3-9 in the previous section shows an AHD of the *SBC_Book* system in which both aggregated systems such as *SBC_Book*, *Part_1*, *Part_2* and non-aggregated systems such as *Chapter_1*, *Chapter_2*, *Chapter_3*,

Chapter_4, *Chapter_5* are displayed. As an interesting contrast, Figure 3-13 in the previous section shows a FD of the *SBC_Book* system in which only non-aggregated systems such as *Chapter_1*, *Chapter_2*, *Chapter_3*, *Chapter_4* and *Chapter_5* are displayed.

3-3 Component Channel Diagram

Systems architects use a component channel diagram (CChD) to display all components' channels of a system. CChD is the third fundamental diagram to achieve structure-behavior coalescence.

3-3-1 Channels of Components

A channel is a model for agent communication via message passing. As shown in Figure 3-14, a message may be sent over a channel, and another agent is able to receive messages sent over a channel it has a reference to.

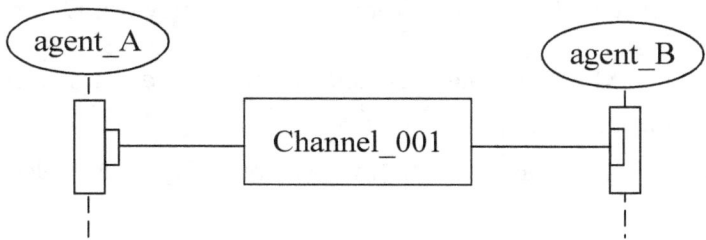

Figure 3-14 A Channel Model

A channel may contain several input parameters (e.g. i_1, i_2) and output parameters (e.g. o_1, o_2), as shown in Figure 3-15.

Figure 3-15 A Channel Contains Several Input/Output Parameters

A channel formula is used to completely describe a channel. A channel formula includes a) channel name, b) input parameters (e.g. $i_1, i_2, ..., i_m$) and c) output parameters (e.g. $o_1, o_2, ..., o_n$), as shown in Figure 3-16.

$$\text{Channel\_Name (In } i_1, i_2, ..., i_m \text{ ; Out } o_1, o_2, ..., o_n)$$

Figure 3-16   Channel Formula

An interaction represents an indivisible and instantaneous communication or handshake between two agents. In the channel-based approach as shown in Figure 3-17, the caller agent (either external environment's actor or component) interacts with the callee agent (component) through the channel interaction.

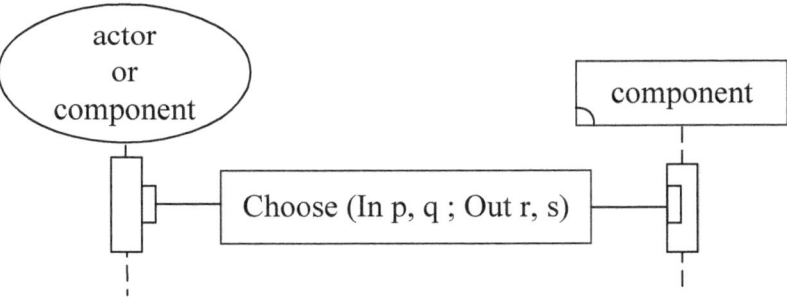

Figure 3-17   Channel-Based Value-Passing Interaction

In order to simplify the channel-based interaction diagram, we will redraw it as shown in Figure 3-18.

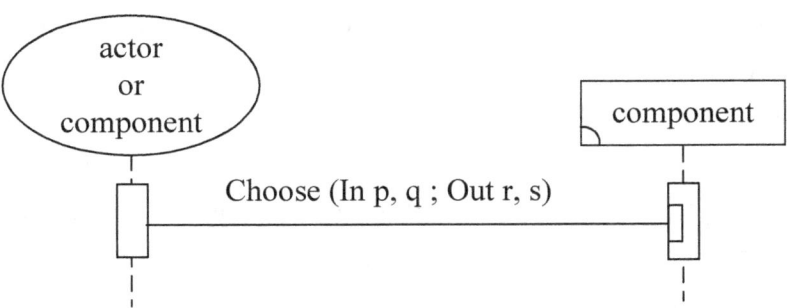

Figure 3-18   Channel-Based Interaction Diagram (I)

Or we can draw the channel-based interaction diagram as shown in Figure 3-19.

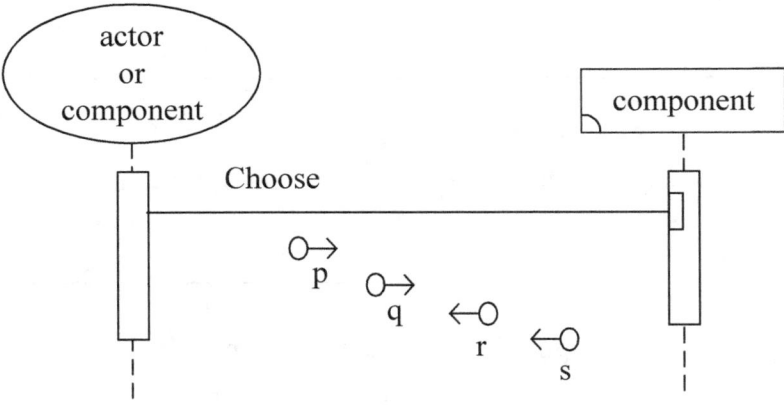

Figure 3-19   Channel-Based Interaction Diagram (II)

In general, a channel is owned by the callee agent (component). Each component in a system must possess at least one channel. A component should not exist in a system if it does not possess any channel. Figure 3-20 shows that component *SalePurchase_UI* has four channels: *SaleInputClick*, *SalePrintClick*, *PurchaseInputClick* and *PurchasePrintClick*.

Figure 3-20   Four Channels of the *SalePurchase_GUI* Component

A channel may have several input and output parameters. The input and output parameters, gathered from all channels, represent the input data and output data views of a system [Date03, Elma10]. As shown in Figure 3-21, component *SalePrint_UI* possesses the *ShowModal* channel which has no input/output parameter; component *SalePrint_UI* also possesses the *SalePrintButtonClick* channel which has the *sDate* and *sNo* input parameters (with the arrow direction pointing to the component) and the *s_report* output parameter (with the arrow direction opposite to the component).

Figure 3-21   Input/Output Parameters of *SalePrintButtonClick*

Data formats of input and output parameters can be described by data type specifications. There are two sets of data types: primitive and composite [Date03, Elma10]. Figure 3-22 shows the primitive data type specification of the *sDate* and *sNo* input parameters occurring in the *SalePrintButtonClick(In sDate, sNo; Out s_report)* channel formula.

Parameter	Data Type	Instances
sDate	Text	20100517, 20100612
sNo	Text	001, 002

Figure 3-22   Primitive Data Type Specification

Figure 3-23 shows the composite data type specification of the *s_report* output parameter occurring in the *SalePrintButtonClick(In sDate, sNo; Out s_report)* channel formula.

Parameter	*s_report*

Data Type	TABLE of
	  Sale Date : Text
	  Sale No : Text
	  Customer : Text
	  ProductNo : Text
	  Quantity : Integer
	  UnitPrice : Real
	  Total : Real
	End TABLE;

Instances	Sale Date : 20100517 Sale No : 001 Customer : Larry Fink | ProductNo | Quantity | UnitPrice | |---|---|---| | A12345 | 400 | 100.00 | | A00001 | 300 | 200.00 | Total : 100,000.00

Figure 3-23   Composite Data Type Specification

3-3-2 Drawing the Component Channel Diagram

For a system, CChD is used to display all components' channels. Figure 3-24 shows the *Multi-Tier Personal Data System's* CChD. In the figure, component *MTPDS_UI* has four channels: *Calculate_AgeClick_Call*, *Calculate_AgeClick_Return*, *Calculate_OverweightClick_Call*, *Calculate_OverweightClick_Return*; component *Age_Logic* has two channels: *Calculate_Age_Call*, *Calculate_Age_Return*; component *Overweight_Logic* has two channels: *Calculate_Overweight_Call*, *Calculate_Overweight_Return*; component *Personal_Database* has two channels: *Sql_DateOfBirth_Select* and *Sql_SexHeightWeight_Select*.

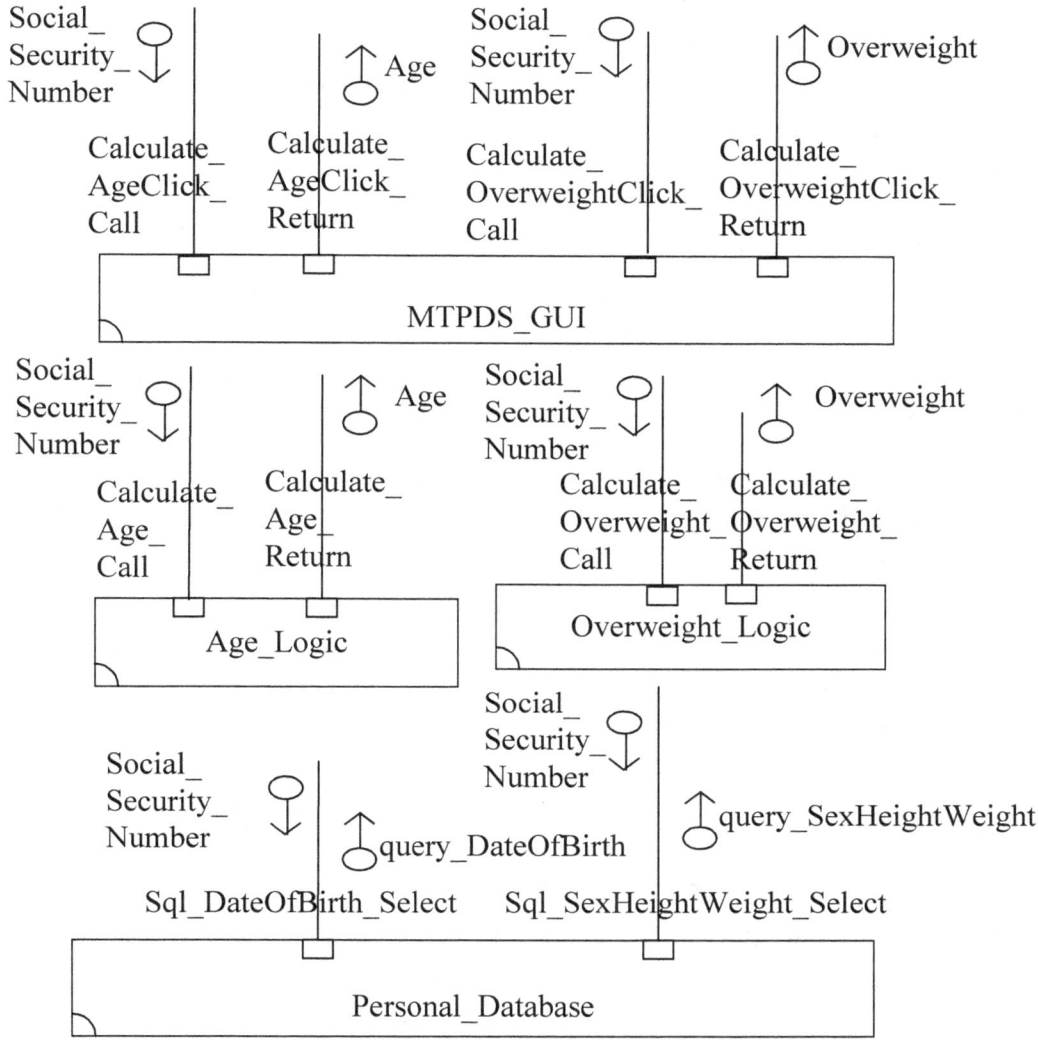

Figure 3-24   CChD of the *Multi-Tier Personal Data System*

The channel formula of *Calculate_AgeClick_Call* is *Calculate_AgeClick_Call(In Social_Security_Number)*. The channel formula of *Calculate_AgeClick_Return* is *Calculate_AgeClick_Return(Out Age)*. The channel formula of *Calculate_OverweightClick_Call* is *Calculate_OverweightClick_Call(In Social_Security_Number)*. The channel formula of *Calculate_OverweightClick_Return* is *Calculate_OverweightClick_Return(Out Overweight)*. The channel formula of *Calculate_Age_Call* is *Calculate_Age_Call(In Social_Security_Number)*. The channel formula of *Calculate_Age_Return* is

Calculate_Age_Return(Out Age). The channel formula of *Calculate_Overweight_Call* is *Calculate_Overweight_Call(In Social_Security_Number)*. The channel formula of *Calculate_Overweight_Return* is *Calculate_Overweight_Return(Out Overweight)*. The channel formula of *Sql_DateOfBirth_Select* is *Sql_DateOfBirth_Select(In Social_Security_Number; Out query_DateOfBirth)*. The channel formula of *Sql_SexHeightWeight_Select* is *Sql_SexHeightWeight_Select(In Social_Security_Number; Out query_SexHeightWeight)*.

Figure 3-25 shows the primitive data type specification of the *Social_Security_Number* input parameter and the *Age*, *Overweight* output parameters.

Parameter	Data Type	Instances
Social_Security_Number	Text	424-87-3651, 512-24-3722
Age	Integer	28, 56
Overweight	Boolean	Yes, No

Figure 3-25   Primitive Data Type Specification

Figure 3-26 shows the composite data type specification of the *query_DateOfBirth* output parameter occurring in the *Sql_DateOfBirth_Select(In Social_Security_Number; Out query_DateOfBirth)* channel formula.

Parameter	*query_DateOfBirth*

Data Type	TABLE of
	   Social_Security_Number : Text
	   Age : Integer
	End TABLE ;

Instances	424-87-3651 28 512-24-3722 56

Figure 3-26   Composite Data Type Specification

Figure 3-27 shows the composite data type specification of the *query_SexHeightWeight* output parameter occurring in the *Sql_SexHeightWeight_Select(In Social_Security_Number; Out query_SexHeightWeight)* channel formula.

Parameter	*query_SexHeightWeight*

Data Type	TABLE of
	   Social_Security_Number : Text
	   Sex : Text
	   Height : Number
	   Weight : Number
	End TABLE ;

Instances	424-87-3651 Female 162 76 512-24-3722 Male 180 80

Figure 3-27   Composite Data Type Specification

3-4 Component Connection Diagram

A component connection diagram (CCoD) is utilized to describe how all components and actors are connected within a system. CCoD is the fourth fundamental diagram to achieve structure-behavior coalescence.

3-4-1 Essence of a Connection

A connection implies a channel request. When a channel is used by another subsystem then a connection appears. Accordingly, a connection is defined as the linkage that is constructed when a channel is used by another subsystem. Figure 3-28 shows that Subs*ystem_A* uses the *Salary_Calculation* channel provided by the *Component_B* component.

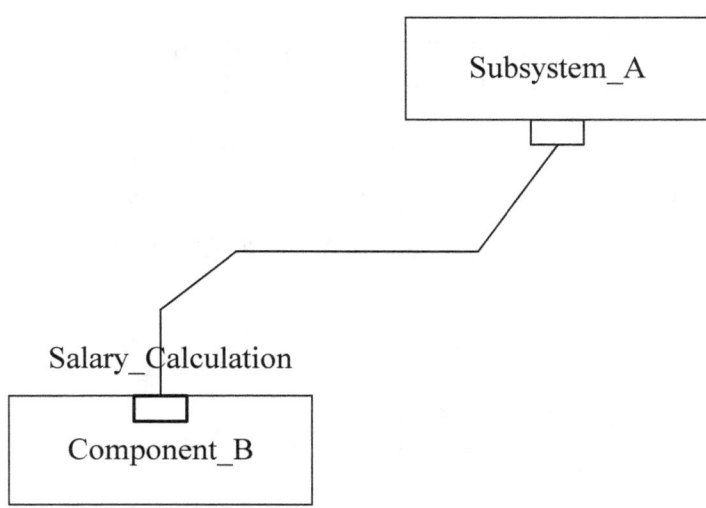

Figure 3-28   A Connection Appears When a Channel is Used

The above figure describes, sufficiently, the essence of a connection. However, we seldom use this kind of drawing. Instead, a simplified drawing of the above figure is often used as shown in Figure 3-29.

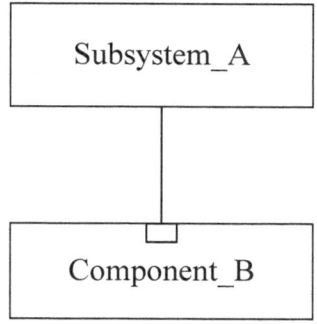

Figure 3-29   Simplified Drawing of a Connection

Since a channel is always provided by a component, there is no doubt that the *Component_B* channel provider is a component. On the contrary, the *Subsystem_A* channel user can be either a component (e.g., *Component_A*) or an actor (e.g., *Actor_A*) as shown in Figure 3-30. An actor belongs to the external environment of a system.

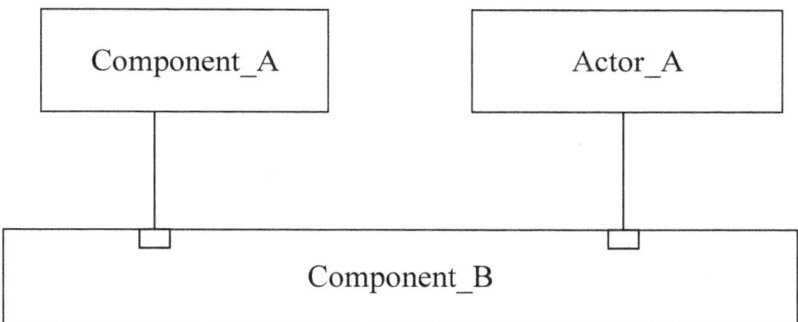

Figure 3-30   Channel User is Either a Component Or an Actor

Within a connection the subsystem (either a component or an actor) using the channel is always entitled the *Client* and the component which provides the channel is always entitled the *Server* as Figure 3-31 shows.

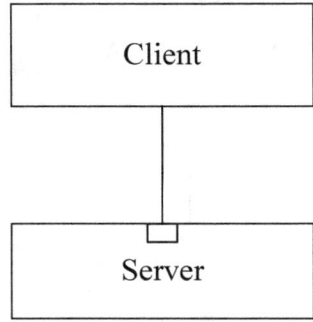

Figure 3-31   Roles of Client and Server Within a Connection

3-4-2 Drawing the Component Connection Diagram

A component connection diagram (CCoD) is utilized to describe how all components and actors (in the external environment) are connected within a system. Figure 3-32 exhibits the *Multi-Tier Personal Data System's* CChD.

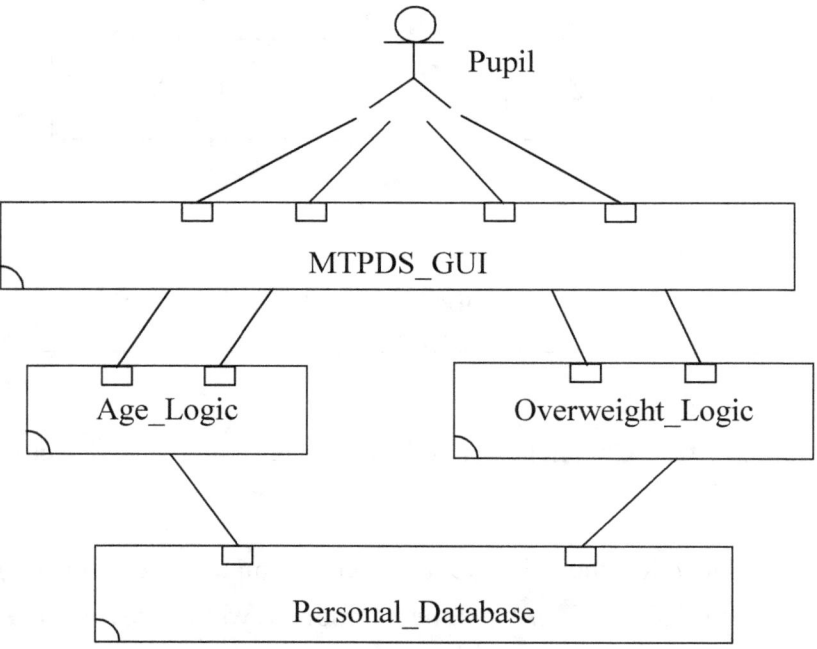

Figure 3-32   CCoD of the *Multi-Tier Personal Data System*

In Figure 3-32, actor *Pupil* has four connections with the *MTPDS_UI* component; component *MTPDS_UI* has two connections with each of the *Age_Logic* and *Overweight_Logic* components; component *Age_Logic* has a connection with the *Personal_Database* component; component *Overweight_Logic* has a connection with the *Personal_Database* component.

After finishing the CCoD, the formation pattern of the *Multi-Tier Personal Data System* will be constructed; thus the systems structure of the *Multi-Tier Personal Data System* becomes more transparent.

Chapter 4: Systems Behavior

SBC-ADL uses the structure-behavior coalescence diagram and interaction flow diagram to delineate the systems behavior of a system.

4-1 Structure-Behavior Coalescence Diagram

Structure-behavior coalescence diagram (SBCD) enables a system architect to observe the structure and behavior coexisting in a system. SBCD is the fifth fundamental diagram to achieve structure-behavior coalescence.

4-1-1 Purpose of Structure-Behavior Coalescence Diagram

The major aim of the SBC-ADL approach is to achieve the integration of systems structure and systems behavior within a system. SBCD enables a system architect to observe the systems structure and systems behavior coexisting in a system. This is the purpose of utilizing SBCD when architecting the systems architecture.

Figure 4-1 exhibits the *Multi-Tier Personal Data System*'s SBCD In this example, interactions among the *Pupil* actor and the *MTPDS_UI*, *Age_Logic*, *Overweight_Logic* and *Personal_Database* components shall draw forth the *AgeCalculation* and *OverweightCalculation* behaviors.

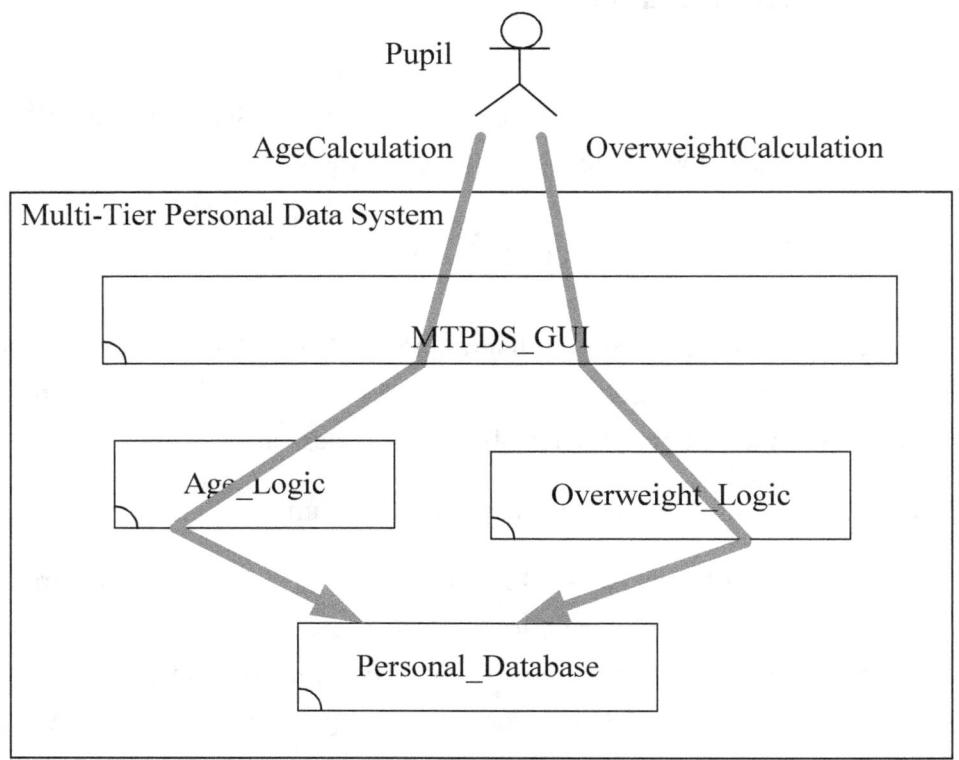

Figure 4-1   SBCD of the *Multi-Tier Personal Data System*

The overall behavior of a system is the aggregation of all its individual behaviors. All individual behaviors are mutually independent of each other. They tend to be executed concurrently [Hoar85, Miln89, Miln99]. For example, the overall behavior of the *Multi-Tier Personal Data System* includes the *AgeCalculation* and *OverweightCalculation* behaviors. In other words, the *AgeCalculation* and *OverweightCalculation* behaviors are combined to produce the overall behavior of the *Multi-Tier Personal Data System*.

The major purpose of using the architectural approach, instead of separating the structure model from the behavior model, is to achieve a coalesced model. In Figure 4-1, systems architects are able to see the systems structure and systems behavior coexisting in a SBCD. That is, in the *Multi-Tier Personal Data System's* SBCD, we not only see its systems structure but also see (at the same time) its systems behavior.

4-1-2 Drawing the Structure-Behavior Coalescence Diagram

Let us now explain the usage of SBCD by constructing a SBCD step by step. The goal of having a SBCD is enabling systems architects to see both the structure and behavior, simultaneously. In order to achieve this goal, a SBCD is drawn by first constructing all of the components, then describing the external environment's actors, and finally describing the interactions among these components and the external environment's actors.

For example, the *Multi-Tier Personal Data System* has two behaviors: *AgeCalculation* and *OverweightCalculation*. After constructing the *Multi-Tier Personal Data System* with all its components, the external environment's actors and the *AgeCalculation* behavior, we obtain the graphical representation as shown in Figure 4-2. In this Figure, the *AgeCalculation* behavior indicates that actor *Pupil* interacts with the *MTPDS_UI* component first, then component *MTPDS_UI* interacts with the *Age_Logic* component later, then component *Age_Logic* interacts with the *Personal_Database* component finally.

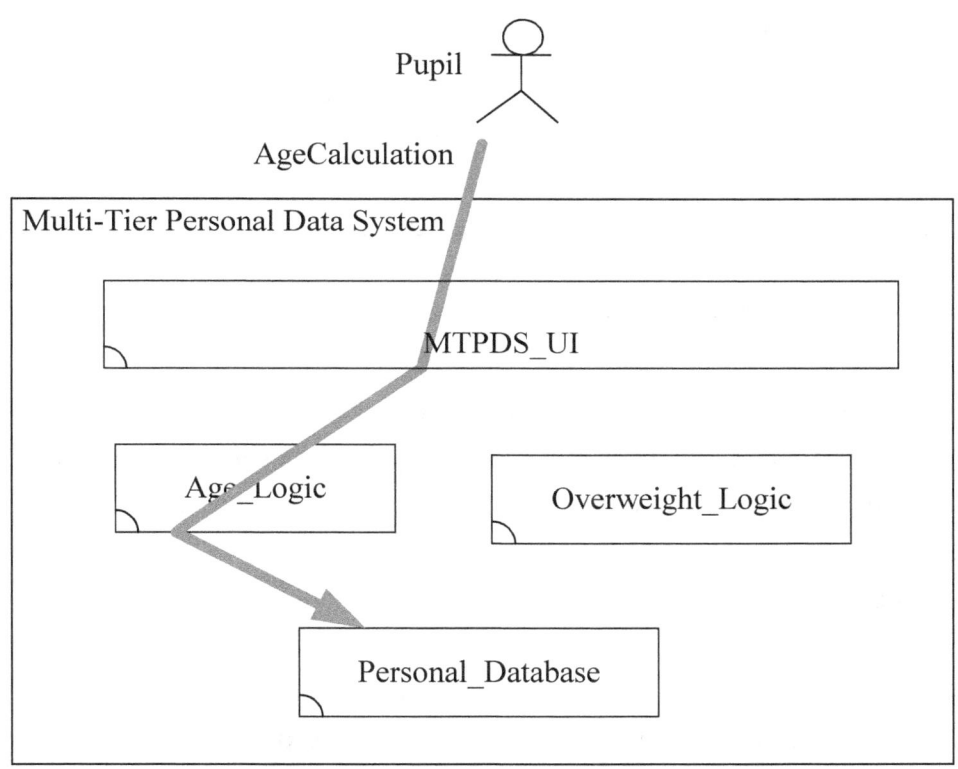

Figure 4-2   All Components, Actors, and the *AgeCalculation* Behavior

Adding the *OverweightCalculation* behavior to Figure 4-2, we then obtain the graphical representation shown in Figure 4-3. In this Figure, the *OverweightCalculation* behavior indicates that actor *Pupil* interacts with the *MTPDS_UI* component first, then component *MTPDS_UI* interacts with the *Overweight_Logic* component later, then component *Overweight_Logic* interacts with the *Personal_Database* component finally.

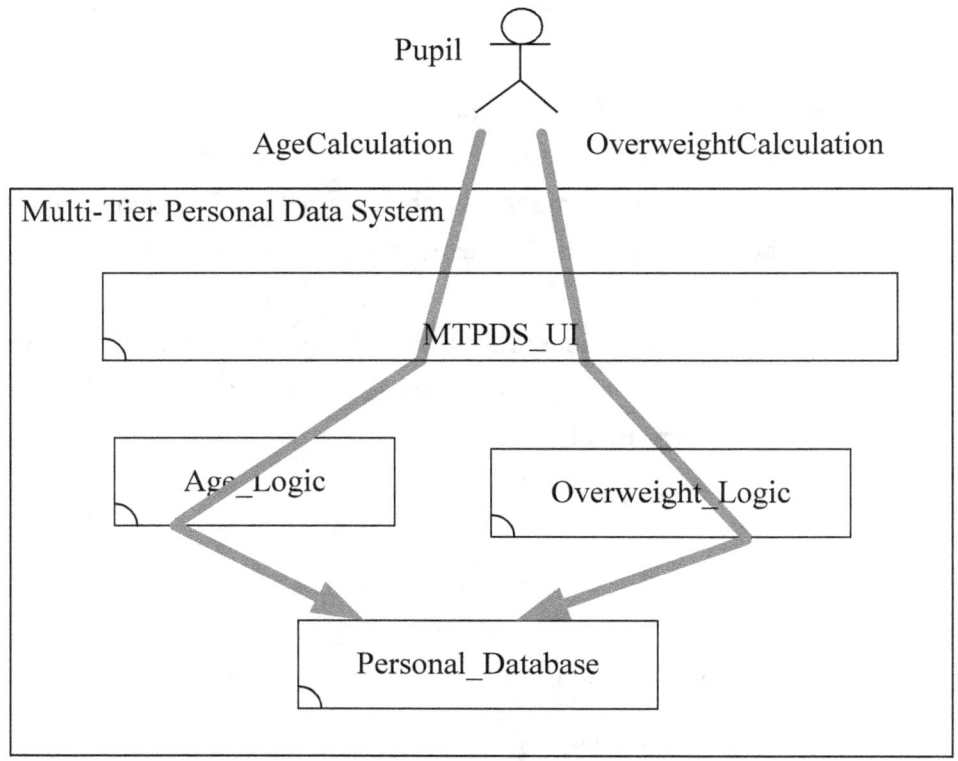

Figure 4-3   Adding the *OverweightCalculation* Behavior to Figure 4-2

After finishing Figure 4-3, we actually have accomplished all the works needed to draw an entire SBCD of the *Multi-Tier Personal Data System*. As a matter of fact, Figure 4-3 shows exactly the *Multi-Tier Personal Data System*'s SBCD.

4-2 Interaction Flow Diagram

An interaction flow diagram (IFD) is utilized to describe each individual behavior of the overall behavior of a system. IFD is the sixth fundamental diagram to achieve structure-behavior coalescence.

4-2-1 Individual Systems Behavior Represented by Interaction Flow Diagram

The overall behavior of a system consists of many individual behaviors. Each individual behavior represents an execution path. An IFD is utilized to represent such an individual behavior.

Figure 4-4 demonstrates that the *Multi-Tier Personal Data System* has two behaviors; thus, it has two IFDs.

Enterprise	IFD
Multi-Tier Personal Data System	AgeCalculation
	OverweightCalculation

Figure 4-4 *Multi-Tier Personal Data System* has Two IFDs

4-2-2 Drawing the Interaction Flow Diagram

Let us now explain the usage of interaction flow diagram (IFD) by drawing an IFD step by step. Figure 4-5 demonstrates an IFD of the *SaleInput* behavior. The X-axis direction is from the left side to right side and the Y-axis direction is from the above to the below. Inside an IFD, there are four elements: a) external environment's actor, b) components, c) interactions and d) input/output parameters. Participants of the interaction, such as the external environment's actor and each component, are laid aside along the X-axis direction on the top of the diagram. The external environment's actor which initiates the sequential interactions is always placed on the most left side of the X-axis. Then, interactions among the external environment's actor and components successively in turn decorate along the Y-axis direction. The first interaction is placed on the top of the Y-axis position. The last interaction is placed on the bottom of the Y-axis position. Each interaction may carry several input and/or output parameters.

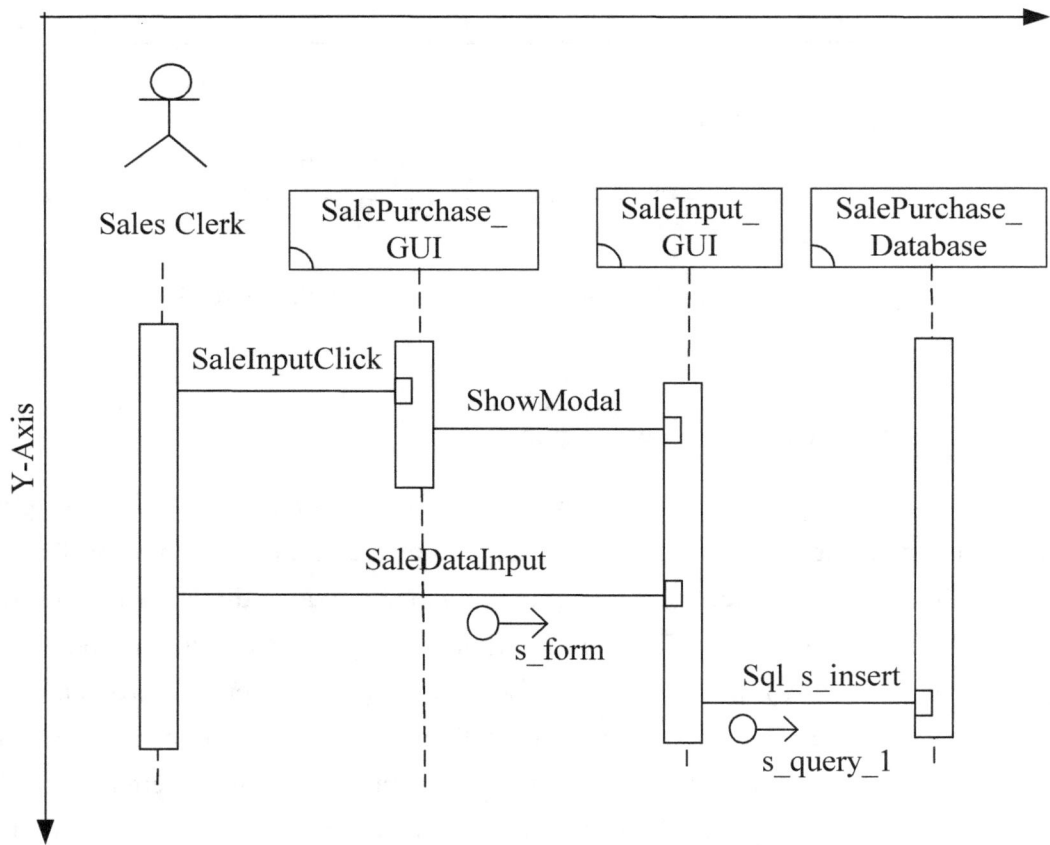

Figure 4-5   IFD of the *SaleInput* Behavior

In Figure 4-5, *Sales Clerk* is an external environment's actor. *SalePurchase_UI*, *SaleInput_UI* and *SalePurchase_Database* are components. *SaleInputClick* is a channel which is provided by the *SalePurchase_UI* component. *ShowModal* is a channel which is provided by the *SaleInput_UI* component. *SaleDataInput* is a channel, carrying the *s_form* input parameter, which is also provided by the *SaleInput_UI* component. *Sql_s_insert* is a channel, carrying the *s_query_1* input parameter, which is provided by the *SalePurchase_Database* component.

The execution path of Figure 4-5 is as follows. First, actor *Sales Clerk* interacts with the *SalePurchase_UI* component through the *SaleInputClick* channel interaction. Next, component *SalePurchase_UI* interacts with the *SaleInput_UI* component through the *ShowModal* channel interaction. Continuingly, actor *Sales Clerk* interacts with the *SaleInput_UI* component through the *SaleDataInput* channel interaction, carrying the *s_form* input parameter. Finally, component *SaleInput_UI* interacts with the *SalePurchase_Database* component through the *Sql_s_insert* channel interaction, carrying the *s_query_1* input parameter.

An interaction flow diagram may contain a conditional expression. Figure 4-6 shows such an example which has the following execution path. First, external environment's actor *Employee* interacts with the *Computer* component through the *Open_Call* channel interaction, carrying the *Task_No* input parameter. Next, if the *var_1 < 4 & var_2 > 7* condition is true then component *Computer* shall interact with the *Skype* component through the *Ch_1_Call* channel interaction and component *Skype* shall interact with the *Earphone* component through the *Ch_4_Call* channel interaction, carrying the *Skype_Earphone* output parameter; else if the *var_3 = 99* condition is true then component *Computer* shall interact with the *Skype* component through the *Ch_2_Call* channel interaction and component *Skype* shall interact with the *Speaker* component through the *Ch_5* channel interaction, carrying the *Skype_Speaker* output parameter; else component *Computer* shall interact with the *Youtube* component through the *Ch_3_Call* channel interaction and component *Youtube* shall interact with the *Speaker* component through the *Ch_6* channel interaction, carrying the *Youtube_Speaker* output parameter. Continuingly, if the *var_1 < 4 & var_2 > 7* condition is true then component *Computer* shall interact with the *Skype* component through the *Ch_1_Return* channel interaction, carrying the *Status_1* output parameter; else if the *var_3 = 99* condition is true then component *Computer* shall interact with the *Skype* component through the *Ch_2_Return* channel interaction, carrying the *Status_2* output parameter; else component *Computer* shall interact with the *Youtube* component through the *Ch_3_Return* channel interaction, carrying the *Status_3* output parameter. Finally, external environment's actor *Employee* interacts with the *Computer* component through the *Open_Return* channel interaction, carrying the *Status* output parameter.

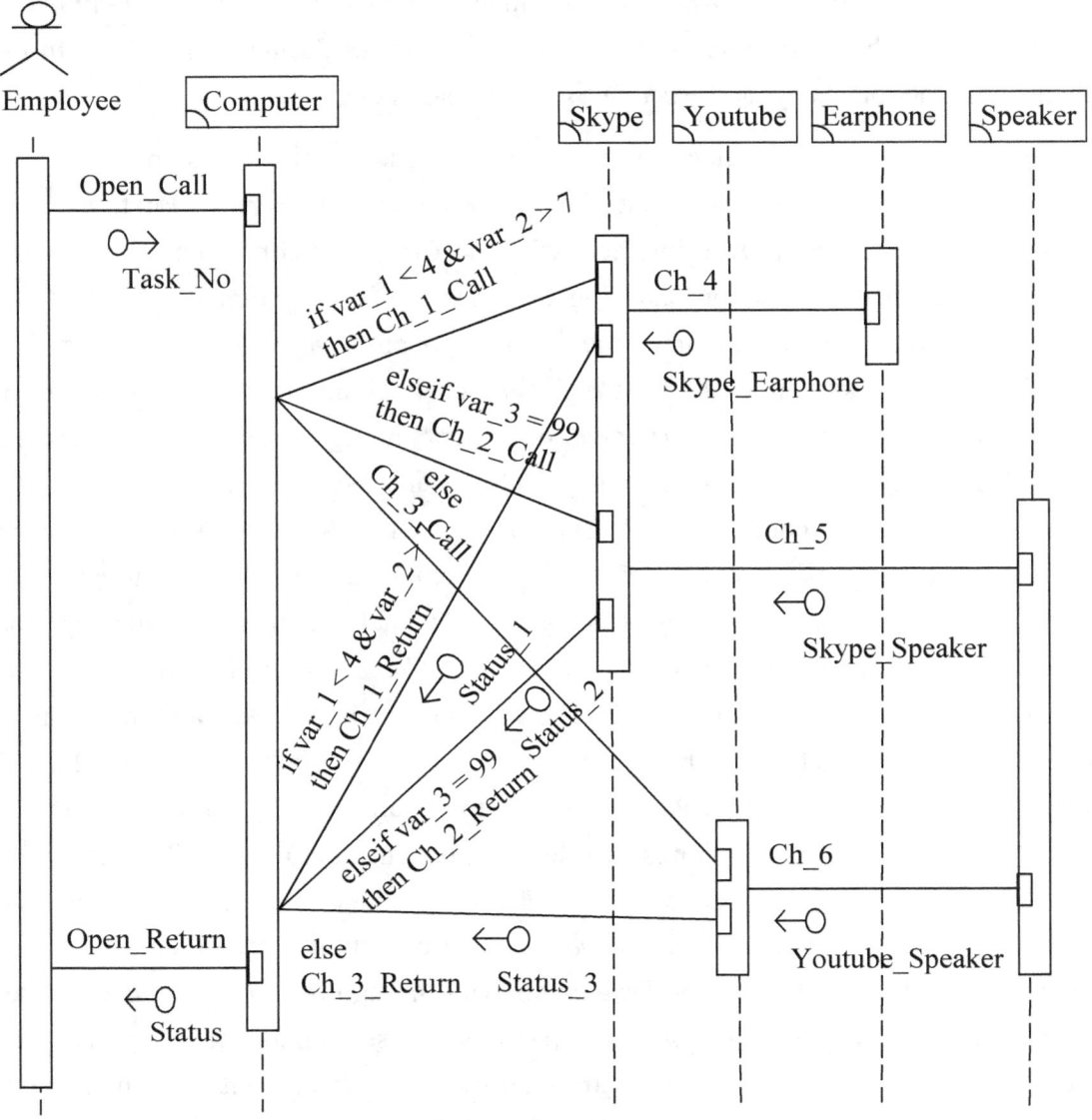

Figure 4-6   Conditional Expression

Several Boolean conditions are shown in Figure 4-6. They are "*var_1 < 4 & var_2 > 7*" and "*var_3 = 99*". Variables, such as *var_1*, *var_2* and *var_3*, appearing in the Boolean condition can be local or global variables [Prat00, Seth96].

PART III: SYSTEMS ARCHITECTURE OF PSCASIS

Chapter 5: AHD of the PSCASIS

AHD is the architecture hierarchy diagram we obtain after the architecture construction is finished. Figure 5-1 shows an AHD of the *Patient Surveillance Cloud Applications and Services IoT System* (PSCASIS). In the figure, *PSCASIS* is composed of *Application_Layer*, *Data_Layer* and *Technology_Layer*; *Application_Layer* is composed of *Presentation_Layer* and *Logic_Layer*; *Presentation_Layer* is composed of *Patient_Account_Registering_UI*, *Alerts_Notifying_UI*, *Emergency_Response_Starting_Time_UI* and *Emergency_Response_End_Time_UI*; *Logic_Layer* is composed of *Patient_Vital_Signs_Daemon*; *Data_Layer* is composed of *PSCASIS_Database*; *Technology_Layer* is composed of *Patient_Vital_Signs_Sensor_P (P = 0000 to 9999)*.

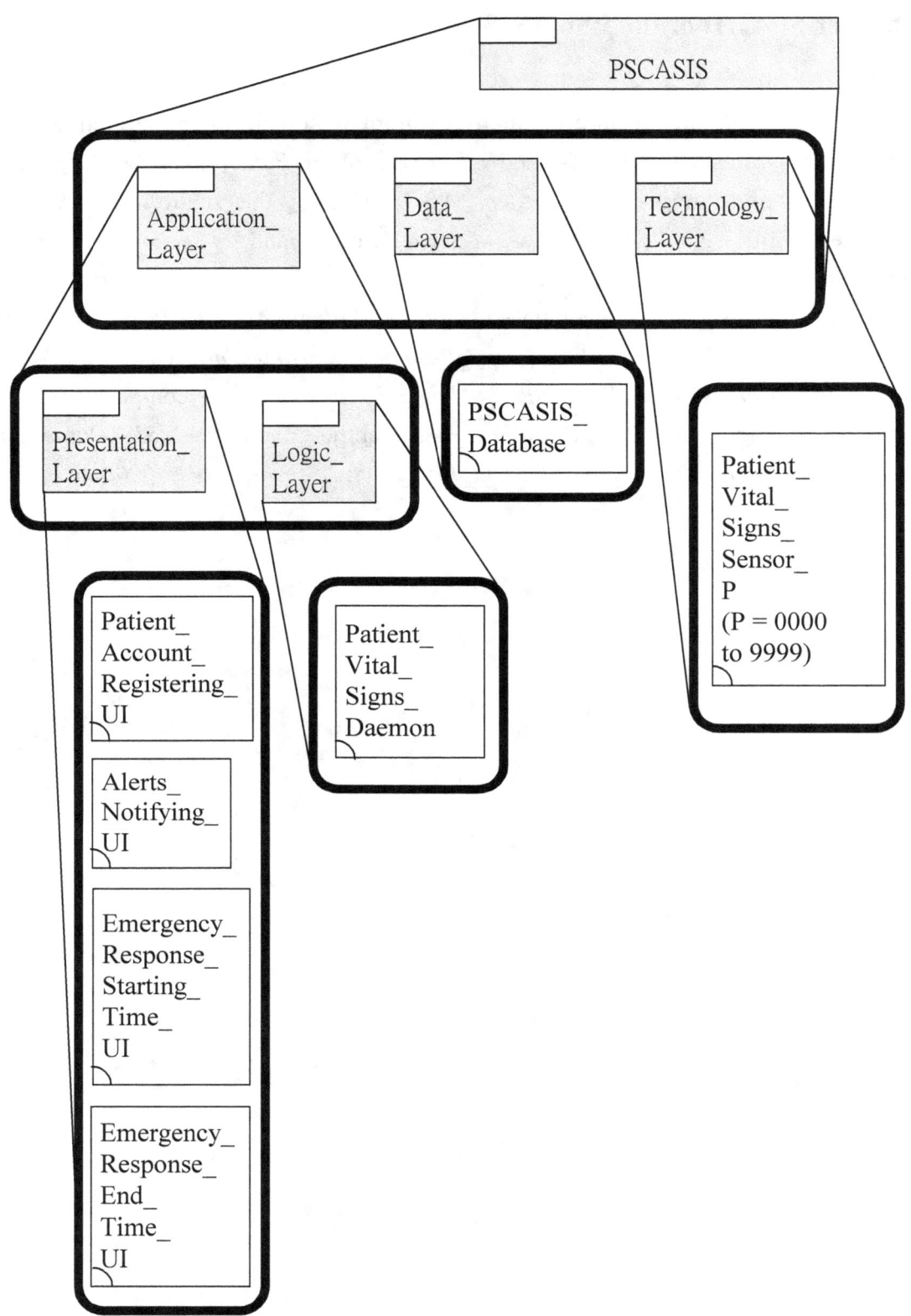

Figure 5-1   AHD of the *PSCASIS*

In Figure 5-1, *PSCASIS, Application_Layer*, *Presentation_Layer*, *Logic_Layer*, *Data_Layer* and *Technology_Layer* are aggregated systems while *Patient_Account_Registering_UI*, *Alerts_Notifying_UI*, *Emergency_Response_Starting_Time_UI*, *Emergency_Response_End_Time_UI*, *Patient_Vital_Signs_Daemon*, *PSCASIS_Database* and *Patient_Vital_Signs_Sensor_P (P = 0000 to 9999)* are non-aggregated systems.

Chapter 6: FD of the PSCASIS

FD is the framework diagram we obtain after the architecture construction is finished. Figure 6-1 shows a FD of the *Patient Surveillance Cloud Applications and Services IoT System* (PSCASIS).

Figure 6-1   FD of the *PSCASIS*

In the above figure, *Presentation_Layer* and *Logic_Layer* are sub-layers of *Application_Layer*. *Presentation_Layer* contains the *Patient_Account_Registering_UI*, *Alerts_Notifying_UI*, *Emergency_Response_Starting_Time_UI* and *Emergency_Response_End_Time_UI* components; *Logic_Layer* contains the *Patient_Vital_Signs_Daemon* component; *Data_Layer* contains the *PSCASIS_Database* component; *Technology_Layer* contains the *Patient_Vital_Signs_Sensor_P (P = 0000 to 9999)* components.

Chapter 7: CChD of the PSCASIS

CChD is the component channel diagram we obtain after the architecture construction is finished. Figure 7-1 shows a CChD of the *Patient Surveillance Cloud Applications and Services IoT System* (PSCASIS). In the figure, component *Patient_Account_Registering_UI* has one channel: *Input_Patient_Account*; component *Alerts_Notifying_UI* has two channels: *Show_All_Alerts* and *Display_Alerts*; component *Emergency_Response_Starting_Time_UI* has one channel: *Input_Emergency_Response_Starting_Time*; component *Emergency_Response_End_Time_UI* has one channel: *Input_Emergency_Response_End_Time*; component *Patient_Vital_Signs_Daemon* has one channel: *Fork_PVSD_Process*; component *PSCASIS_Database* has five channels: *SQL_Insert_Patient_Account*, *SQL_Insert_Patient_Vital_Signs*, *SQL_Select_Patient_Vital_Signs_for_Alerts_Analysis*, *SQL_Insert_Emergency_Response_Starting_Time* and *SQL_Insert_Emergency_Response_End_Time*; component *Patient_Vital_Signs_Sensor_P (P = 0000 to 9999)* has two channels: *Sense_Patient_Vital_Signs* and *Return_Patient_Vital_Signs*.

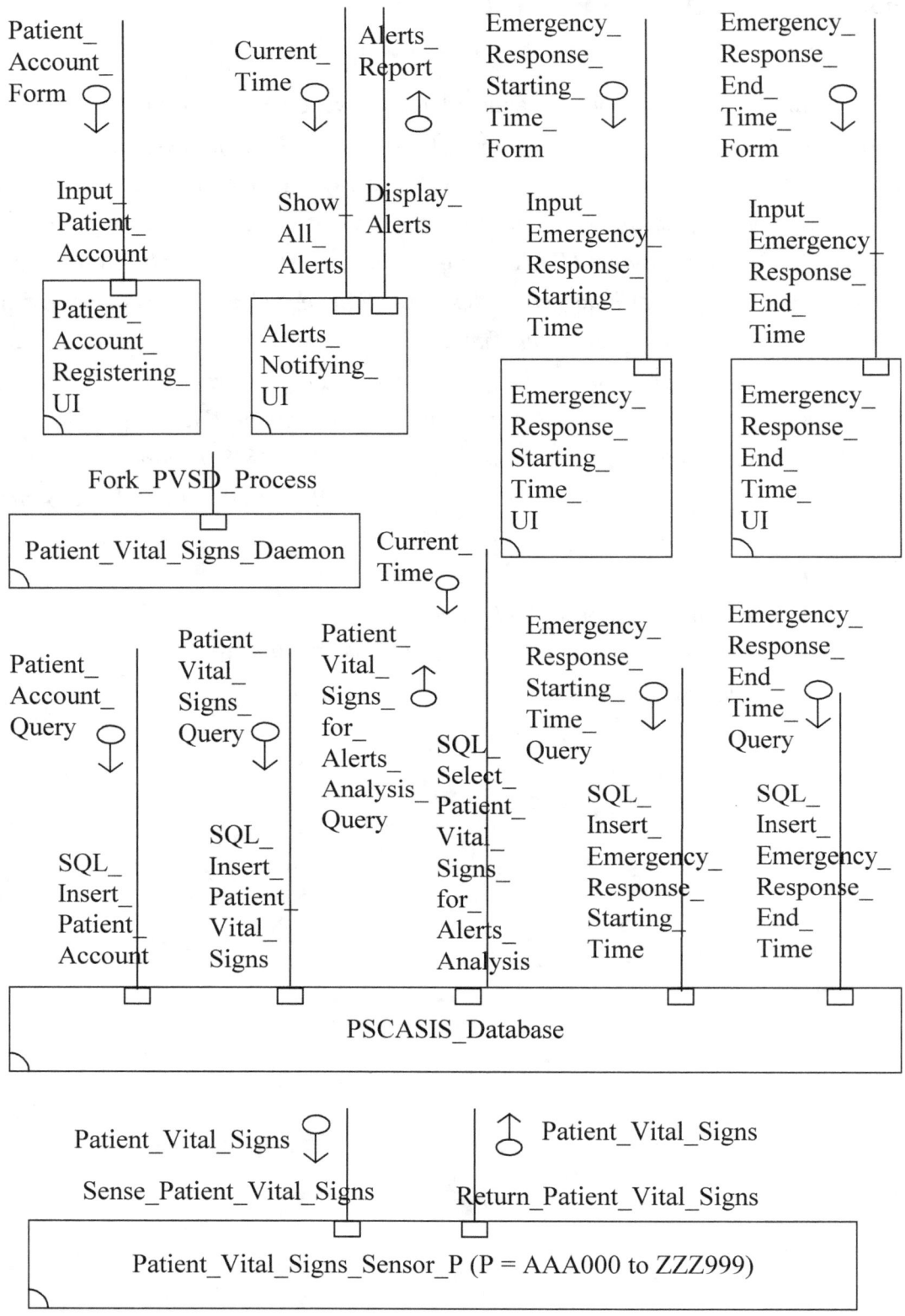

Figure 7-1   CChD of the *PSCASIS*

The channel formula of *Input_Patient_Account* is *Input_Patient_Account(In Patient_Account_Form)*. The channel formula of *Fork_PVSD_Process* is *Fork_PVSD_Process*. The channel formula of *Show_All_Alerts* is *Show_All_Alerts(In Current_Time)*. The channel formula of *Display_Alerts* is *Display_Alerts(Out Alerts_Report)*. The channel formula of *Input_Emergency_Response_Starting_Time* is *Input_Emergency_Response_Starting_Time(In Emergency_Response_Starting_Time_Form)*. The channel formula of *Input_Emergency_Response_End_Time* is *Input_Emergency_Response_End_Time(In Emergency_Response_End_Time_Form)*. The channel formula of *SQL_Insert_Patient_Account* is *SQL_Insert_Patient_Account(In Patient_Account_Query)*. The channel formula of *SQL_Insert_Patient_Vital_Signs* is *SQL_Insert_Patient_Vital_Signs(In Patient_Vital_Signs_Query)*. The channel formula of *SQL_Select_Patient_Vital_Signs_for_Alerts_Analysis* is *SQL_Select_Patient_Vital_Signs_for_Alerts_Analysis(In Current_Time; Out Patient_Vital_Signs_for_Alerts_Analysis_Query)*. The channel formula of *SQL_Insert_Emergency_Response_Starting_Time* is *SQL_Insert_Emergency_Response_Starting_Time(In Emergency_Response_Starting_Time_Query)*. The channel formula of *SQL_Insert_Emergency_Response_End_Time* is *SQL_Insert_Emergency_Response_End_Time(In Emergency_Response_End_Time_Query)*. The channel formula of *Sense_Patient_Vital_Signs* is *Sense_Patient_Vital_Signs(In Patient_Vital_Signs)*. The channel formula of *Return_Patient_Vital_Signs* is *Return_Patient_Vital_Signs(Out Patient_Vital_Signs)*.

Figure 7-2 shows the composite data type specification of the *Patient_Account_Form* input parameter occurring in the *Input_Patient_Account(In Patient_Account_Form)* channel formula.

	Patient_Account_Form

Data Type	TABLE of
	  Patient_Number: Text
	  Name: Text
	  Sex: Text
	  Birthday: Text
	  Address: Text
	  Phone: Text
	  Bed_Number: Text
	End TABLE ;

Instances	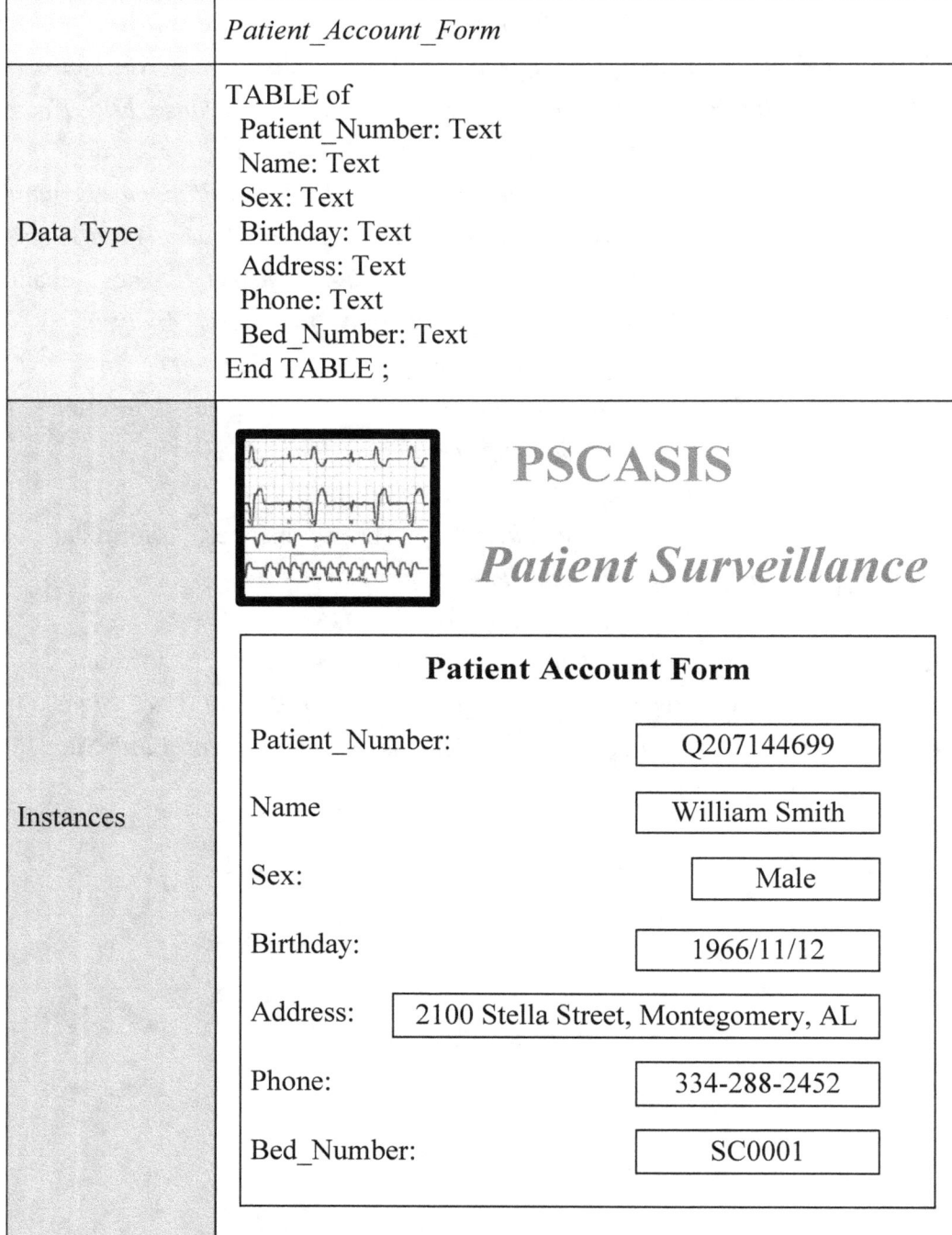

Figure 7-2   Composite Data Type Specification of *Patient_Account_Form*

Figure 7-3 shows the primitive data type specification of the *Current_Time* parameter occurring in the *Show_All_Alerts(In Current_Time)* and *SQL_Select_Patient_Vital_Signs_for_Alerts_Analysis(In Current_Time; Out Patient_Vital_Signs_for_Alerts_Analysis_Query)* channel formulas.

Parameter	Data Type	Instances
Current_Time	Text	2017/08/15, 14:25

Figure 7-3   Primitive Data Type Specification

Figure 7-4 shows the composite data type specification of the *Alerts_Report* output parameter occurring in the *Display_Alerts(Out Alerts_Report)* channel formula.

Parameter	*Alerts_Report*

Data Type	TABLE of
	   Current_Time: Text
	   Patient_Number: Text
	   Name: Text
	   Alert_Code: Text
	   Emergency_Response: Boolean
	End TABLE ;

Instances	**PSCASIS** *Patient Surveillance* **Alerts Report** 2017/08/15, 14:25PM | Patient_Number	Name	Alert_Code	Emergency_Response | |---|---|---|---| | A207144699 | Grace Trump | 10000 | YES | | B411277322 | John Richardson | 10000 | NOT YET | | B125670055 | Tom Nixon | 00010 | NOT YET | | C243112288 | Lee Shelton | 00011 | YES |

Figure 7-4   Composite Data Type Specification of *Alerts_Report*

Figure 7-5 shows the composite data type specification of the *Emergency_Response_Starting_Time_Form* input parameter occurring in the *Input_Emergency_Response_Starting_Time(In Emergency_Response_Starting_Time_Form)* channel formula.

Parameter	*Emergency_Response_Starting_Time_Form*

Data Type	TABLE of
	  Patient_Number: Text
	  Name: Text
	  Emergency_Response_Starting_Time: Text
	  Actions: Text
	End TABLE ;

Instances	**PSCASIS** — *Patient Surveillance* **Emergency Response Starting Time** Patient_Number: C243112288 Name: Lee Shelton Emergency_Response_Starting_Time: 2017/08/15, 14:20 Actions: Send a nurse over; Inform the doctor.

Figure 7-5   Composite Data Type Specification of *Emergency_Response_Starting_Time_Form*

Figure 7-6 shows the composite data type specification of the *Emergency_Response_End_Time_Form* input parameter occurring in the *Input_Emergency_Response_End_Time(In Emergency_Response_End_Time_Form)* channel formula.

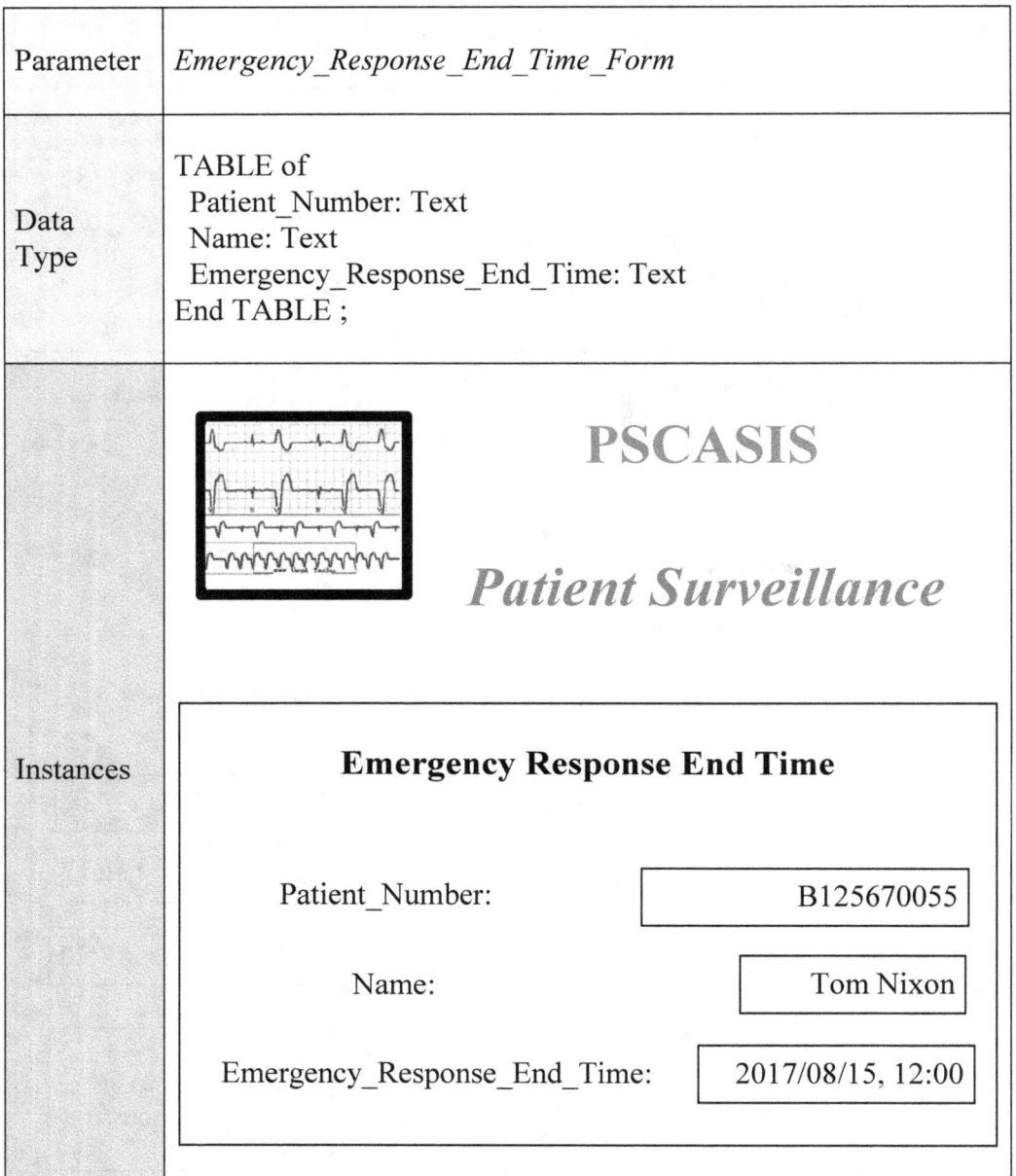

Figure 7-6   Composite Data Type Specification of
Emergency_Response_End_Time_Form

Figure 7-7 shows the composite data type specification of the *Patient_Account_Query* input parameter occurring in the *SQL_Insert_Patient_Account(In Patient_Account_Query)* channel formula.

	Patient_Account_Query

Data Type	TABLE of
	  Patient_Number: Text
	  Name: Text
	  Sex: Text
	  Birthday: Text
	  Address: Text
	  Phone:
	  Bed_Number: Text
	End TABLE ;

Instances	<table><tr><td>Patient_Number</td><td>Q207144699</td></tr><tr><td>Name</td><td>William Smith</td></tr><tr><td>Sex</td><td>Male</td></tr><tr><td>Birthday</td><td>1966/11/12</td></tr><tr><td>Address</td><td>2100 Stella Street, Montegomery, AL</td></tr><tr><td>Phone</td><td>334-288-2452</td></tr><tr><td>Bed_Number</td><td>SC0001</td></tr></table>

Figure 7-7   Composite Data Type Specification of *Patient_Account_Query*

Figure 7-8 shows the composite data type specification of the *Patient_Vital_Signs_Query* input parameter occurring in the *SQL_Insert_Patient_Vital_Signs(In Patient_Vital_Signs_Query)* channel formula.

Parameter	*Patient_Vital_Signs_Query*

Data Type	TABLE of
	  Patient_Number: Text
	  Sensing_Time: Text
	  Heart_Beat: Integer
	  Systolic_ Arterial_ Pressure: Integer
	  Diastolic_Pressure: Integer
	  Pulmonary-Artery_Pressure: Integer/Integer
	  Blood_ Temperature: Real
	End TABLE ;

Instances	<table><tr><th>Patient_Number</th><th>Sensing_Time</th></tr><tr><td>Q207144699</td><td>2017/08/15, 12:20</td></tr></table> <table><tr><th>Heart_ Beat</th><th>Systolic_ Arterial_ Pressure</th><th>Diastolic_ Pressure</th><th>Pulmonary-Artery_ Pressure</th><th>Blood_ Temperature</th></tr><tr><td>82</td><td>120</td><td>67</td><td>30/12</td><td>37.22</td></tr></table>

Figure 7-8   Composite Data Type Specification of *Patient_Vital_Signs_Query*

Figure 7-9 shows the composite data type specification of the *Patient_Vital_Signs_for_Alerts_Analysis_Query* output parameter occurring in the *SQL_Select_Patient_Vital_Signs_for_Alerts_Analysis(In Current_Time; Out Patient_Vital_Signs_for_Alerts_Analysis_Query)* channel formula.

Parameter	*Patient_Vital_Signs_for_Alerts_Analysis_Query*

Data Type	TABLE of
	  Patient_Number: Text
	  Heart_Beat: Integer
	  Systolic_ Arterial_ Pressure: Integer
	  Diastolic_Pressure: Integer
	  Pulmonary-Artery_Pressure: Integer/Integer
	  Blood_ Temperature: Real
	  Emergency_Response_Starting_Time: Text
	End TABLE ;

Instances	<table><tr><th>Patient_ Number</th><th>Heart_ Beat</th><th>Systolic_ Arterial_ Pressure</th><th>Diastolic_ Pressure</th></tr><tr><td>A207144699</td><td>96</td><td>121</td><td>70</td></tr><tr><td>B411277322</td><td>100</td><td>119</td><td>65</td></tr><tr><td>B125670055</td><td>77</td><td>120</td><td>71</td></tr><tr><td>C243112288</td><td>82</td><td>123</td><td>68</td></tr></table> <table><tr><th>Pulmonary-Artery_ Pressure</th><th>Blood_ Temperature</th><th>Emergency_ Response_ Starting_ Time</th></tr><tr><td>29/11</td><td>37.44</td><td>2017/08/15, 13:25</td></tr><tr><td>30/10</td><td>37.12</td><td>NULL</td></tr><tr><td>40/20</td><td>37.04</td><td>NULL</td></tr><tr><td>50/30</td><td>42.55</td><td>2017/08/15, 14:20</td></tr></table>

Figure 7-9 Composite Data Type Specification of *Patient_Vital_Signs_for_Alerts_Analysis_Query*

Figure 7-10 shows the composite data type specification of the *Emergency_Response_Starting_Time_Query* input parameter occurring in the *SQL_Insert_Emergency_Response_Starting_Time(In Emergency_Response_Starting_Time_Query)* channel formula.

Parameter	*Emergency_Response_Starting_Time_Query*

Data Type	TABLE of
	  Patient_Number: Text
	  Name: Text
	  Emergency_Response_Starting_Time: Text
	  Actions: Text
	End TABLE ;

Instances	<table><tr><th>Patient_ Number</th><th>Name</th><th>Emergency_ Response_ Starting_ Time</th><th>Actions</th></tr><tr><td>C243112288</td><td>Lee Shelton</td><td>2017/08/15, 14:20</td><td>Send a nurse over; Inform the doctor.</td></tr></table>

Figure 7-10   Composite Data Type Specification of *Emergency_Response_Starting_Time_Query*

Figure 7-11 shows the composite data type specification of the *Emergency_Response_End_Time_Query* input parameter occurring in the *SQL_Insert_Emergency_Response_End_Time(In Emergency_Response_End_Time_Query)* channel formula.

Parameter	*Emergency_Response_End_Time_Query*

Data Type	TABLE of
	  Patient_Number: Text
	  Name: Text
	  Emergency_Response_End_Time: Text
	End TABLE ;

Instances	<table><tr><th>Patient_Number</th><th>Name</th><th>Emergency_Response_End_Time</th></tr><tr><td>C243112288</td><td>Lee Shelton</td><td>2017/08/15, 15:20</td></tr></table>

Figure 7-11   Composite Data Type Specification of *Emergency_Response_End_Time_Query*

Figure 7-12 shows the composite data type specification of the *Patient_Vital_Signs* parameter occurring in the *Sense_Patient_Vital_Signs(In Patient_Vital_Signs)* and *Return_Patient_Vital_Signs(Out Patient_Vital_Signs)* channel formulas.

Parameter	*Patient_Vital_Signs*				

Data Type	TABLE of
	  Heart_Beat: Integer
	  Systolic_ Arterial_ Pressure: Integer
	  Diastolic_Pressure: Integer
	  Pulmonary-Artery_Pressure: Integer/Integer
	  Blood_ Temperature: Real
	End TABLE ;				

Instances	Heart_ Beat	Systolic_ Arterial_ Pressure	Diastolic_ Pressure	Pulmonary- Artery_ Pressure	Blood_ Temperature
	82	120	67	30/12	37.22

Figure 7-12 Composite Data Type Specification of *Patient_Vital_Signs*

Chapter 8: CCoD of the PSCASIS

CCoD is the component connection diagram we obtain after the architecture construction is finished. Figure 8-1 shows a CCoD of the *Patient Surveillance Cloud Applications and Services IoT System* (PSCASIS).

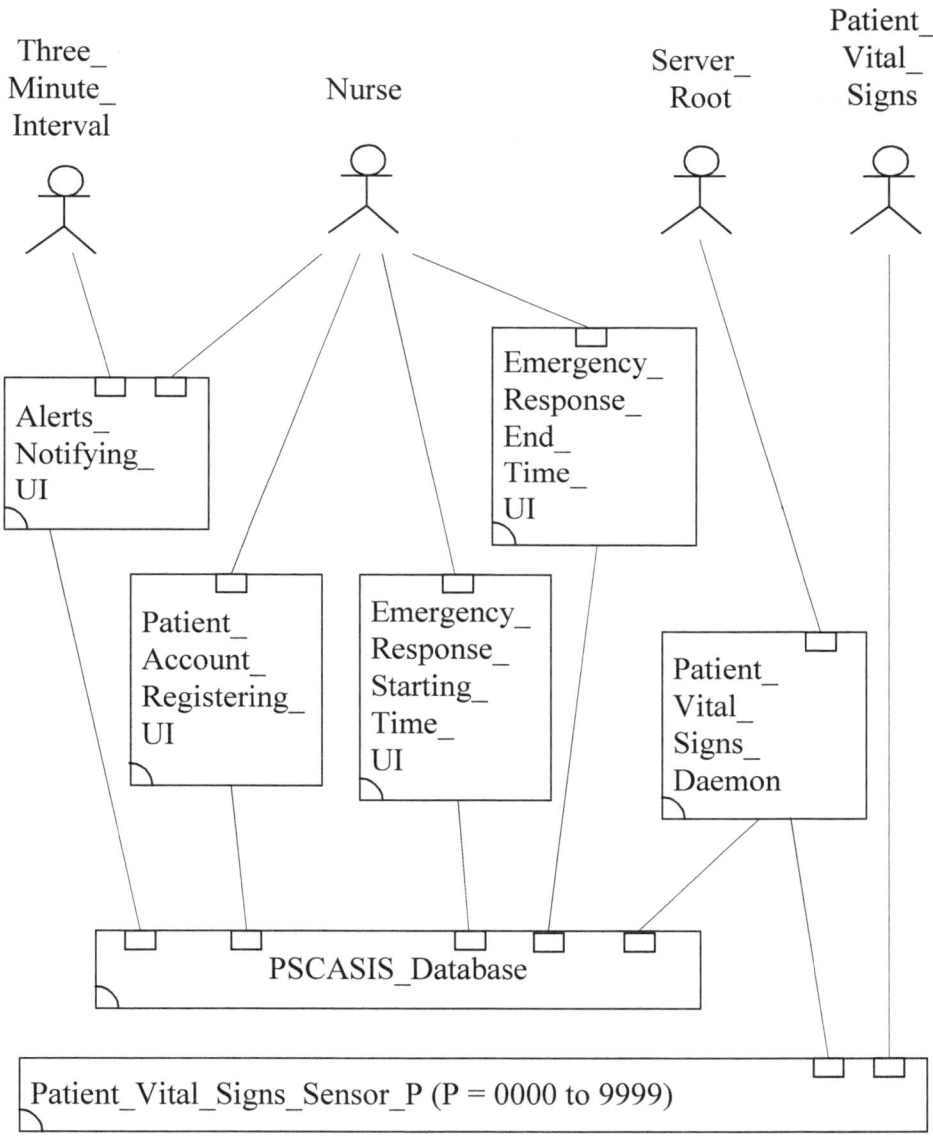

Figure 8-1   CCoD of the *PSCASIS*

In the above figure, actor *Three_Minute_Interval* has a connection with the *Alerts_Notifying_UI* component; actor *Nurse* has a connection with each one of the *Alerts_Notifying_UI*, *Patient_Account_Registering_UI*, *Emergency_Response_Starting_Time_UI*, *Emergency_Response_End_Time_UI* components; actor *Server_Root* has a connection with the *Patient_Vital_Signs_Daemon* component; actor *Patient_Vital_Signs* has a connection with the *Patient_Vital_Signs_Sensor_P (P = 0000 to 9999)* component; each one of the *Alerts_Notifying_UI*, *Patient_Account_Registering_UI*, *Emergency_Response_Starting_Time_UI*, *Emergency_Response_End_Time_UI*, *Patient_Vital_Signs_Daemon* components has a connection with the *PSCASIS_Database* component; component *Patient_Vital_Signs_Daemon* has a connection with the *Patient_Vital_Signs_Sensor_P (P = 0000 to 9999)* component.

Chapter 9: SBCD of the PSCASIS

SBCD is the structure-behavior coalescence diagram we obtain after the architecture construction is finished. Figure 9-1 shows a SBCD of the *Patient Surveillance Cloud Applications and Services IoT System* (PSCASIS) in which interactions among the *Three_Minute_Interval*, *Nurse*, *Server_Root*, *Patient_Vital_Signs* actors and the *Patient_Account_Registering_UI*, *Alerts_Notifying_UI*, *Emergency_Response_Starting_Time_UI*, *Emergency_Response_End_Time_UI*, *Patient_Vital_Signs_Daemon*, *PSCASIS_Database*, *Patient_Vital_Signs_Sensor_P (P = 0000 to 9999)* components shall draw forth the *Registering_Patient_Account*, *Sensing_Patient_Vital_Signs*, *Alerts_Notifying*, *Recording_Emergency_Response_Starting_Time*, *Recording_Emergency_Response_End_Time* behaviors.

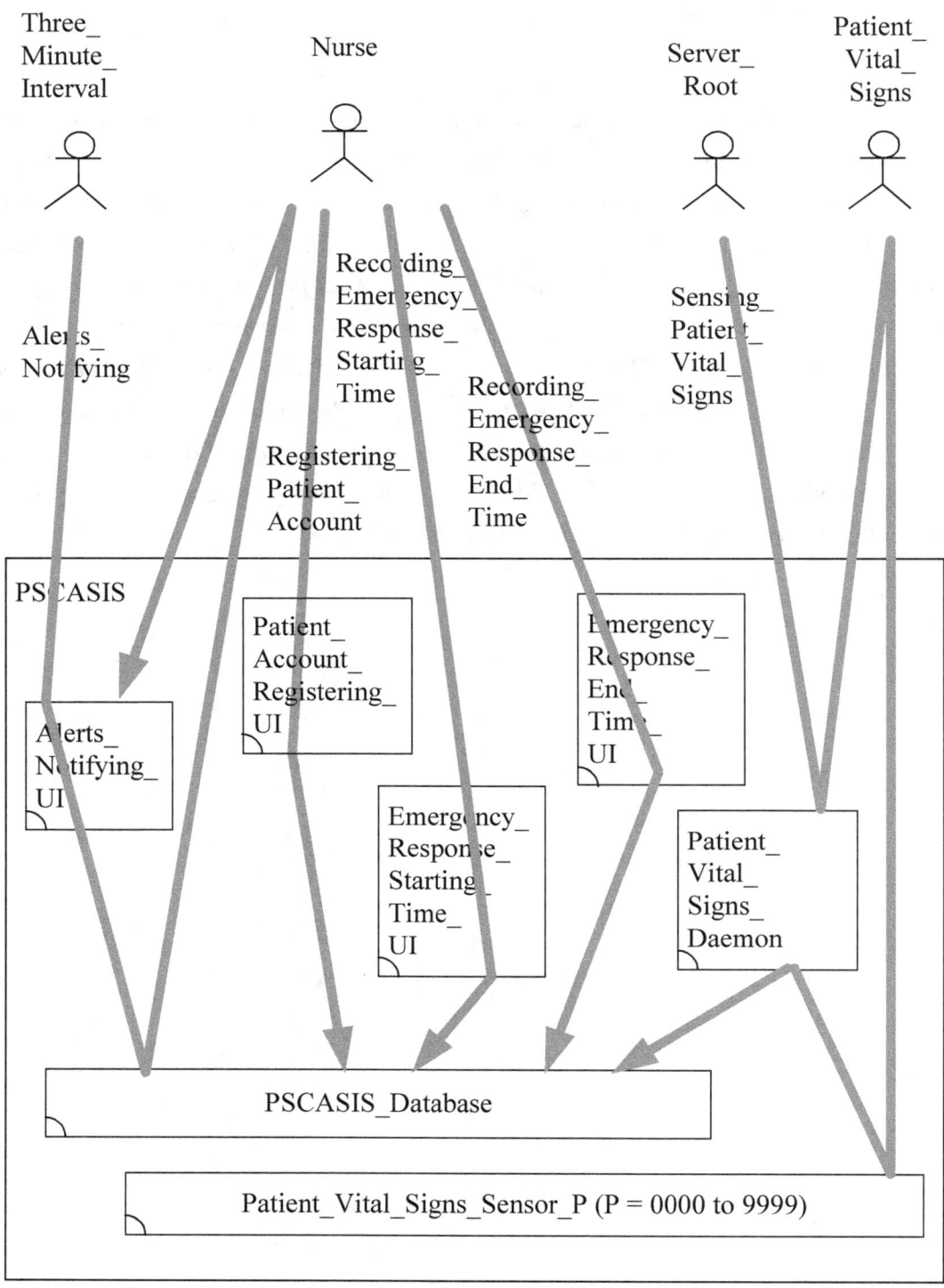

Figure 9-1 SBCD of the *PSCASIS*

The overall behavior of the *Patient Surveillance Cloud Applications and Services IoT System* (PSCASIS) includes the *Registering_Patient_Account*, *Sensing_Patient_Vital_Signs*, *Alerts_Notifying*, *Recording_Emergency_Response_Starting_Time*, *Recording_Emergency_Response_End_Time* behaviors. In other words, the *Registering_Patient_Account*, *Sensing_Patient_Vital_Signs*, *Alerts_Notifying*, *Recording_Emergency_Response_Starting_Time*, *Recording_Emergency_Response_End_Time* behaviors together provide the overall behavior of the *Patient Surveillance Cloud Applications and Services IoT System*.

The major purpose of adopting the architectural approach, instead of separating the structure model from the behavior model, is to achieve one single coalesced model. In Figure 9-1, systems architects are able to see that the systems structure and systems behavior coexist in the SBCD. That is, in the SBCD of *Patient Surveillance Cloud Applications and Services IoT System* (PSCASIS), systems architects not only see its systems structure but also see (at the same time) its systems behavior.

Chapter 10: IFD of the PSCASIS

IFDs are the interaction flow diagrams we obtain after the architecture construction is finished. The overall behavior of the *Patient Surveillance Cloud Applications and Services IoT System* (PSCASIS) includes five individual behaviors: *Registering_Patient_Account*, *Sensing_Patient_Vital_Signs*, *Alerts_Notifying*, *Recording_Emergency_Response_Starting_Time*, *Recording_Emergency_Response_End_Time*. Each individual behavior is represented by an execution path. We use an IFD to define each one of these execution paths.

Figure 10-1 shows an IFD of the *Registering_Patient_Account* behavior. First, actor *Nurse* interacts with the *Patient_Account_Registering_UI* component through the *Input_Patient_Account* channel interaction, carrying the *Patient_Account_Form* input parameter. Finally, component *Patient_Account_UI* interacts with the *PSCASIS_Database* component through the *SQL_Insert_Patient_Account* channel interaction, carrying the *Patient_Account_Query* input parameter.

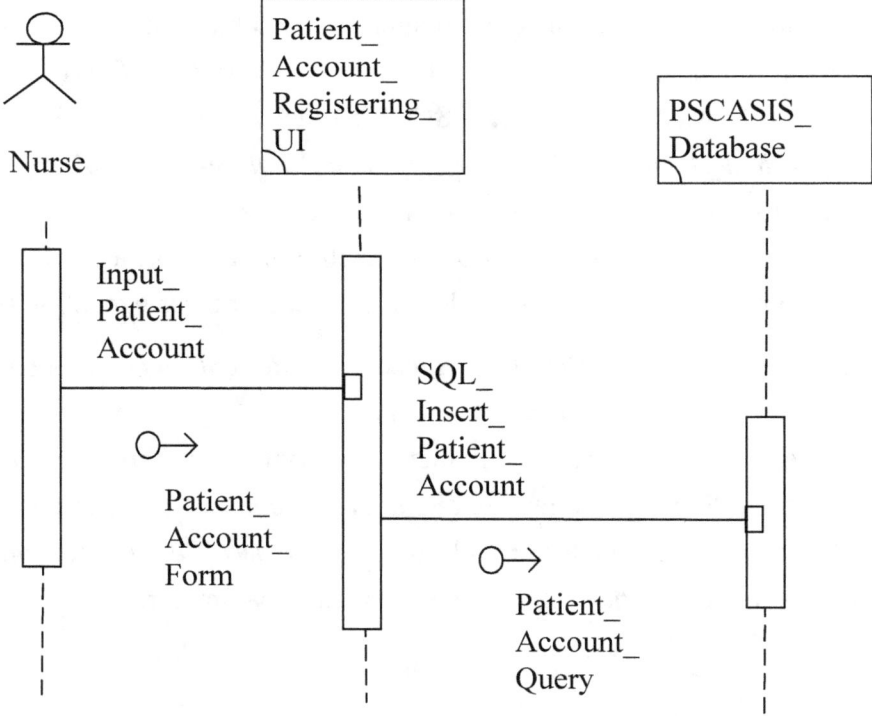

Figure 10-1 IFD of the *Registering_Patient_Account* Behavior

Figure 10-2 shows an IFD of the *Sensing_Patient_Vital_Signs* behavior. First, actor *Server_Root* interacts with the *Patient_Vital_Signs_Daemon* component through the *Fork_PVSD_Process* channel interaction. Next, actor *Patient_Vital_Signs* interacts with the *Patient_Vital_Signs_Sensor_P (P = 0000 to 9999)* component through the *Sense_Patient_Vital_Signs* channel interaction, carrying the *Patient_Vital_Signs* input parameter. Continuingly, component *Patient_Vital_Signs_Daemon* interacts with the *Patient_Vital_Signs_Sensor_P (P = 0000 to 9999)* component through the *Return_Patient_Vital_Signs* channel interaction, carrying the *Patient_Vital_Signs* output parameter. Finally, component *Patient_Vital_Signs_Daemon* interacts with the *PSCASIS_Database* component through the *SQL_Insert_Patient_Vital_Signs* channel interaction, carrying the *Patient_Vital_Signs_Query* input parameter.

Figure 10-2 IFD of the *Sensing_Patient_Vital_Signs* Behavior

Figure 10-3 shows an IFD of the *Alerts_Notifying* behavior. First, actor *Three_Minute_Interval* interacts with the *Alerts_Notifying_UI* component through the *Showing_All_Alerts* channel interaction, carrying the *Current_Time* input parameter. Next, component *Alerts_Notifying_UI* interacts with the *PSCASIS_Database* component through the *SQL_Select_Patient_Vital_Signs_for_Alerts_Analysis* channel interaction, carrying the *Current_Time* input parameter and the *Patient_Vital_Signs_for_Alerts_Analysis_Query* output parameter. Finally, actor *Nurse* interacts with the *Alerts_Notifying_UI* component through the *Display_Alerts* channel interaction, carrying the *Alerts_AreaMap* output parameter.

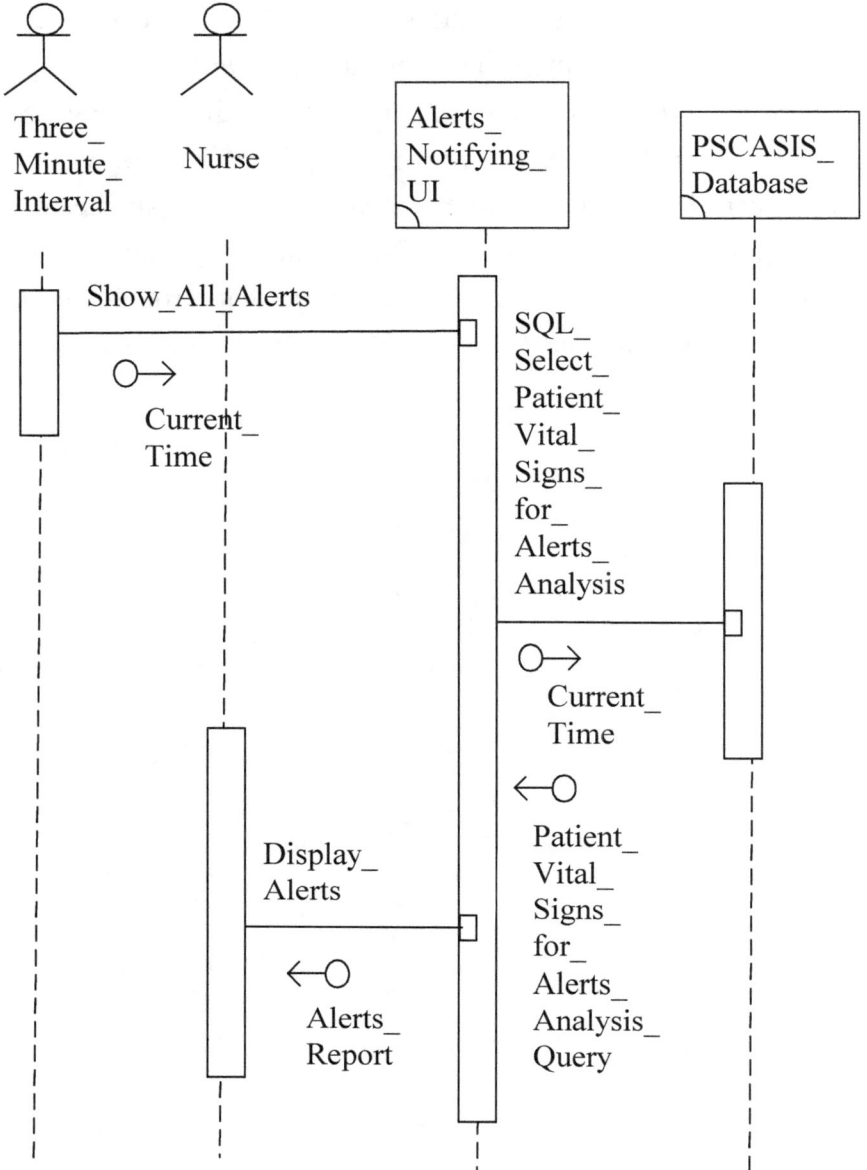

Figure 10-3   IFD of the *Alerts_Notifying* Behavior

Figure 10-4 shows an IFD of the *Recording_Emergency_Response_Starting_Time* behavior. First, actor *Nurse* interacts with the *Emergency_Response_Starting_Time_UI* component through the *Input_Emergency_Response_Starting_Time* channel interaction, carrying the *Emergency_Response_Starting_Time_Form* input parameter. Last, component *Emergency_Response_Starting_Time_UI* interacts with the *PSCASIS_Database* component through the *SQL_Insert_Emergency_Response_Starting_Time* channel interaction, carrying the *Emergency_Response_Starting_Time_Query* input parameter.

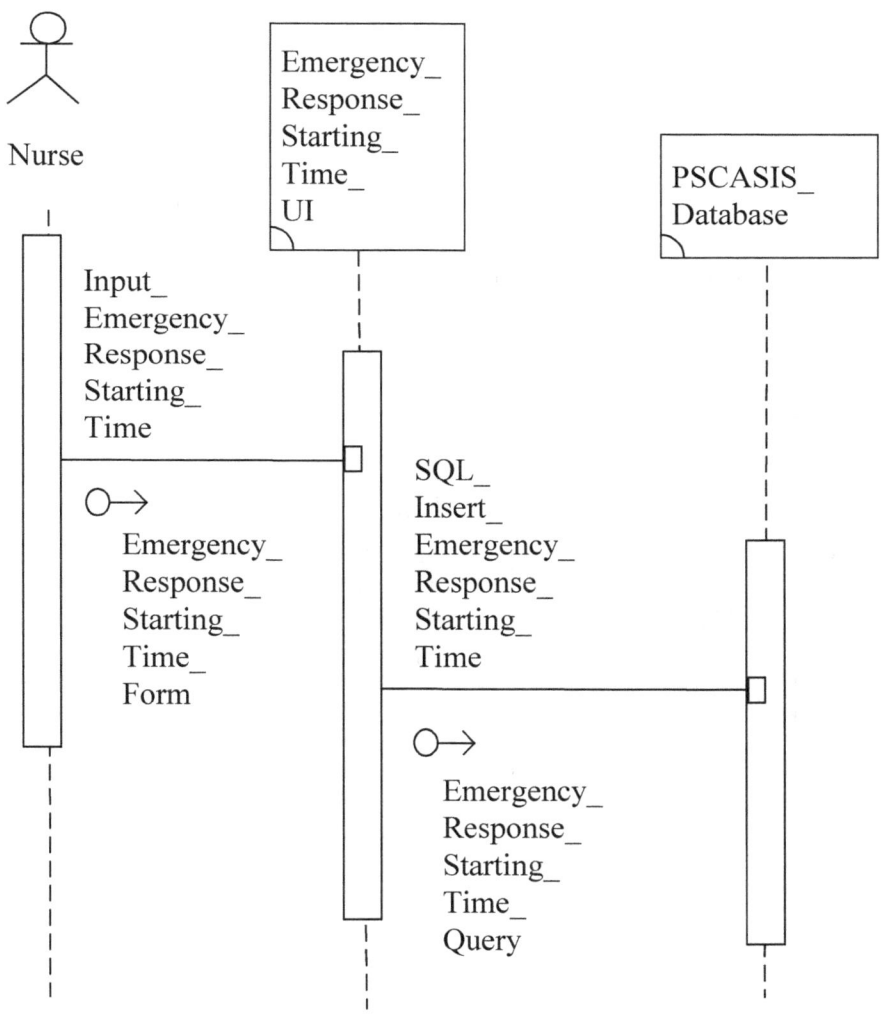

Figure 10-4   IFD of the *Recording_Emergency_Response_Starting_Time* Behavior

Figure 10-5 shows an IFD of the *Recording_Emergency_Response_End_Time* behavior. First, actor *Nurse* interacts with the *Emergency_Response_End_Time_UI* component through the *Input_Emergency_Response_End_Time* channel interaction, carrying the *Emergency_Response_End_Time_Form* input parameter. Last, component *Emergency_Response_End_Time_UI* interacts with the *PSCASIS_Database* component through the *SQL_Insert_Emergency_Response_End_Time* channel interaction, carrying the *Emergency_Response_End_Time_Query* input parameter.

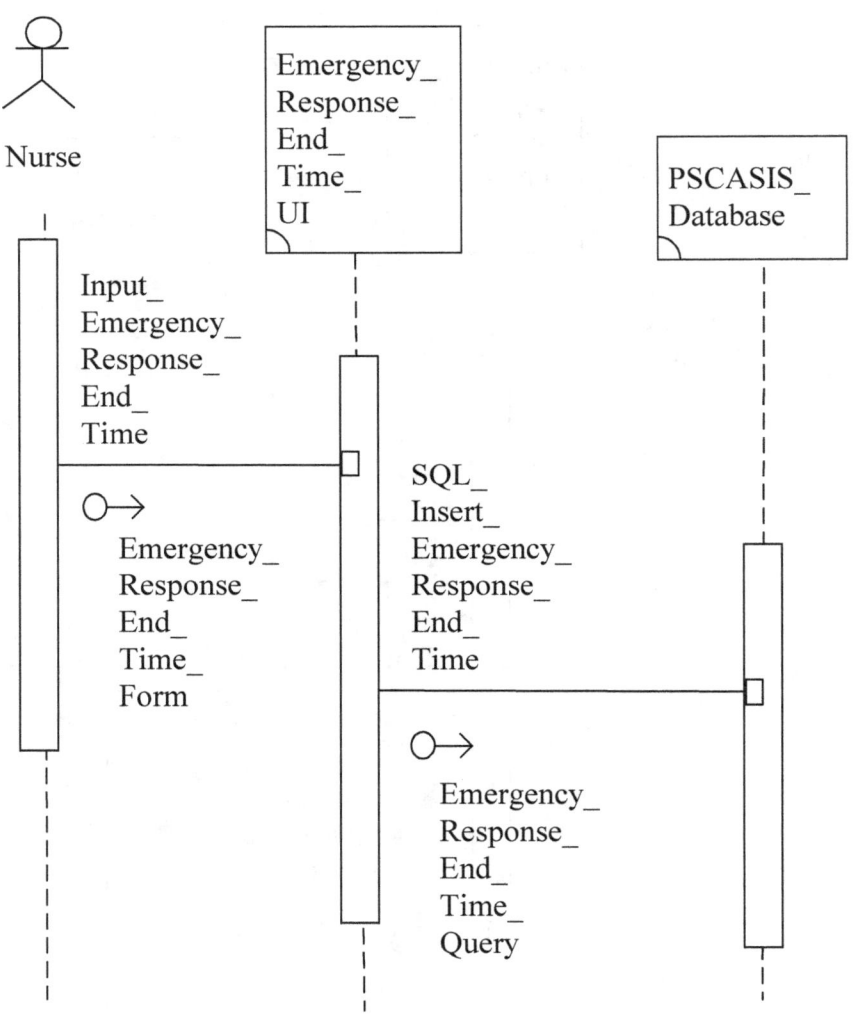

Figure 10-5   IFD of the *Recording_Emergency_Response_End_Time* Behavior

APPENDIX A: SBC ARCHITECTURE DESCRIPTION LANGUAGE

(1) Architecture Hierarchy Diagram

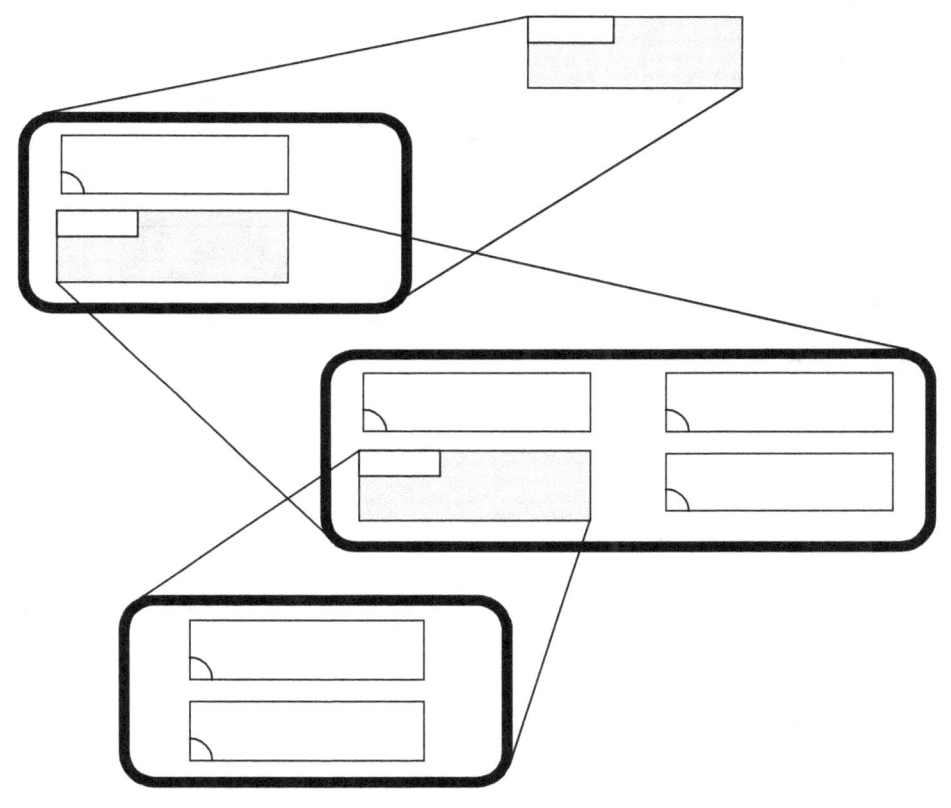

: Aggregated System

: Non-Aggregated System, Component

(2) Framework Diagram

```
┌─────────────────────────────────────────────┐
│ Application Layer                           │
│  ┌───────────────────────────────────────┐  │
│  │ Presentation_Layer                    │  │
│  │              ┌─────────┐  ┌─────────┐ │  │
│  │              └─────────┘  └─────────┘ │  │
│  │                                       │  │
│  │ Logic_Layer                           │  │
│  │              ┌─────────┐  ┌─────────┐ │  │
│  │              └─────────┘  └─────────┘ │  │
│  └───────────────────────────────────────┘  │
│                                             │
│ Data Layer                                  │
│   ┌─────────┐  ┌─────────┐  ┌─────────┐    │
│   └─────────┘  └─────────┘  └─────────┘    │
│                                             │
│ Technology Layer                            │
│     ┌─────────┐    ┌─────────┐              │
│     └─────────┘    └─────────┘              │
└─────────────────────────────────────────────┘
```

───

┌─────────────┐
└─────────────┘   :   Component

(3) Component Channel Diagram

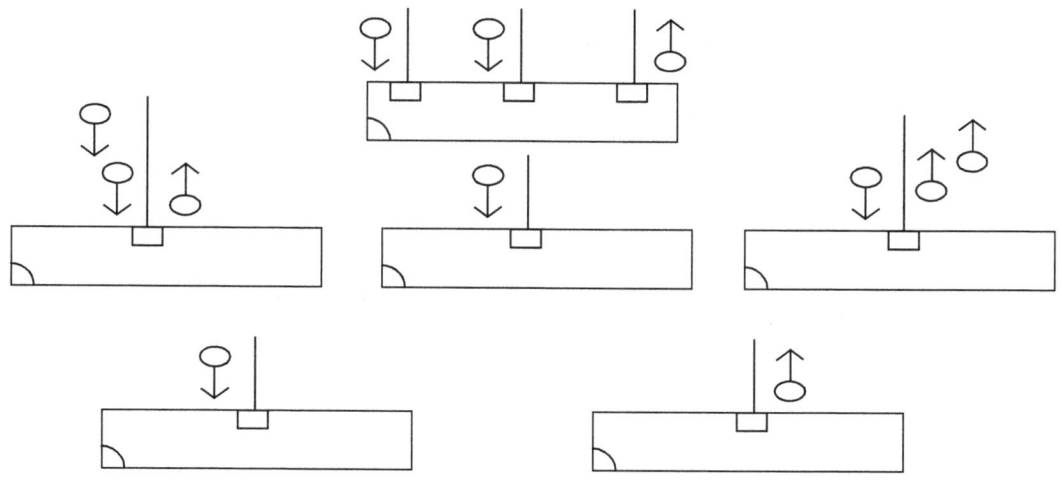

⊥▢	: Channel
⊽	: Input Data
⊼	: Output Data
▭	: Component

(4) Component Connection Diagram

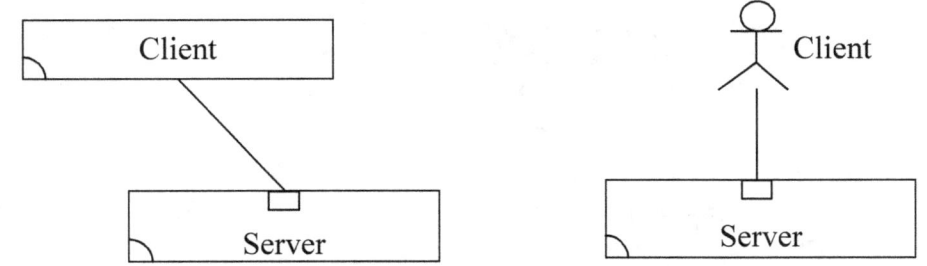

(5) Structure-Behavior Coalescence Diagram

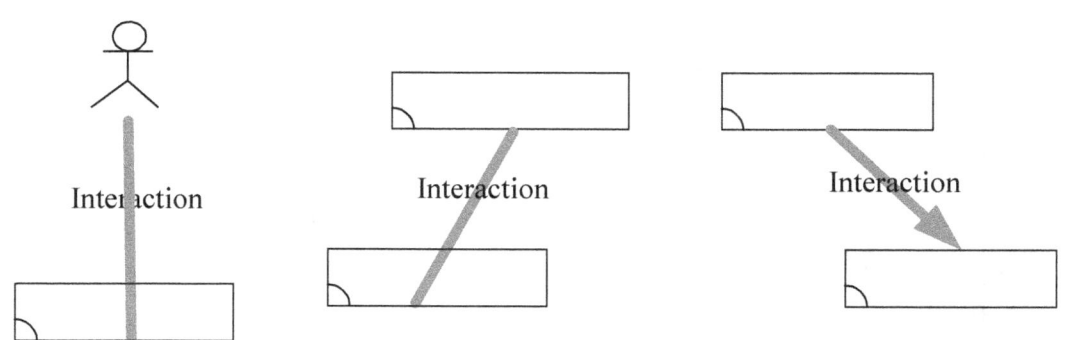

(6) Interaction Flow Diagram

: Channel Interaction

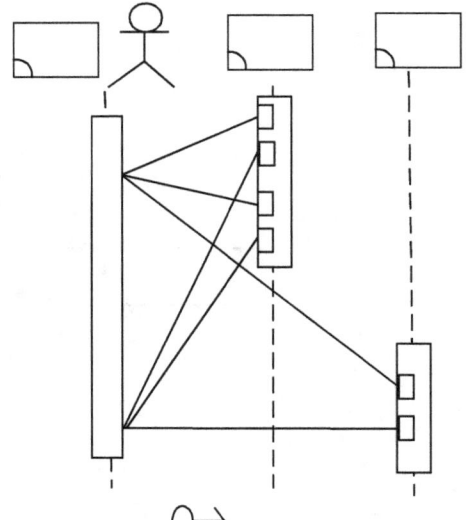
: Conditional Channel Interaction

: Input Data

: Output Data

APPENDIX B: SBC PROCESS ALGEBRA

(1) Channel-Based Single-Queue SBC Process Algebra

(1) <System> ::= "**fix**(" <Process_Variable> "="<IFD> " • " <Process_Variable> {"+" <IFD> " • " <Process_Variable>} ")"

(2) <IFD> ::= <Type_1_Interaction> {" • " <Type_1_Or_2_Interaction>}

(3) <Type_1_Or_2_Interaction> ::= <Type_1_Interaction>

                     | <Type_2_Interaction>

(2) Channel-Based Multi-Queue SBC Process Algebra

(1) <System> ::= <FixIFD> {"||" <FixIFD>}

(2) <FixIFD> ::= **fix**(" <Process_Variable>"="<IFD> "•" <Process_Variable> ")"

(3) <IFD> ::= <Type_1_Interaction> {"•" Type_1_Or_2_Interaction>}

(4) <Type_1_Or_2_Interaction> ::= <Type_1_Interaction>
    | <Type_2_Interaction>

(3) Channel-Based Infinite-Queue SBC Process Algebra

(1) <System> ::= "! ("<IFD> " • " *STOP* ")" {"|| ! (" <IFD> " • " *STOP* ")"}

(2) <IFD> ::= <Type_1_Interaction> {"•" <Type_1_Or_2_Interaction>}

(3) <Type_1_Or_2_Interaction> ::= <Type_1_Interaction>

                                 | <Type_2_Interaction>

BIBLIOGRAPHY

[Bern09] Bernstein, D. et al., "Blueprint for the Intercloud – Protocols and Formats for Cloud Computing Interoperability," *IEEE Computer Society*, 2009, pp.328-336.

[Burd10] Burd, S. D., *Systems Architecture*, 6th Edition, Cengage Learning, 2010.

[Chao14a] Chao, W. S., *Systems Thingking 2.0: Architectural Thinking Using the SBC Architecture Description Language*, CreateSpace Independent Publishing Platform, 2014.

[Chao14b] Chao, W. S., *General Systems Theory 2.0: General Architectural Theory Using the SBC Architecture*, CreateSpace Independent Publishing Platform, 2014.

[Chao14c] Chao, W. S., *Systems Modeling and Architecting: Structure-Behavior Coalescence for Systems Architecture*, CreateSpace Independent Publishing Platform, 2014.

[Chao15a] Chao, W. S., *A Process Algebra For Systems Architecture: The Structure-Behavior Coalescence Approach*, CreateSpace Independent Publishing Platform, 2015.

[Chao15b] Chao, W. S., *An Observation Congruence Model For Systems Architecture: The Structure-Behavior Coalescence Approach*, CreateSpace Independent Publishing Platform, 2015.

[Chao16] Chao, W. S., *System: Contemporary Concept, Definition, and Language*, CreateSpace Independent Publishing Platform, 2016.

[Chao17a] Chao, W. S., *Channel-Based Single-Queue SBC Process Algebra For Systems Definition: General Architectural Theory at Work*, CreateSpace

Independent Publishing Platform, 2017.

[Chao17b] Chao, W. S., *Channel-Based Multi-Queue SBC Process Algebra For Systems Definition: General Architectural Theory at Work*, CreateSpace Independent Publishing Platform, 2017.

[Chao17c] Chao, W. S., *Channel-Based Infinite-Queue SBC Process Algebra For Systems Definition: General Architectural Theory at Work*, CreateSpace Independent Publishing Platform, 2017.

[Chao17d] Chao, W. S., *Operation-Based Single-Queue SBC Process Algebra For Systems Definition: General Architectural Theory at Work*, CreateSpace Independent Publishing Platform, 2017.

[Chao17e] Chao, W. S., *Operation-Based Multi-Queue SBC Process Algebra For Systems Definition: Unification of Systems Structure and Systems Behavior*, CreateSpace Independent Publishing Platform, 2017.

[Chao17f] Chao, W. S., *Operation-Based Infinite-Queue SBC Process Algebra For Systems Definition: Unification of Systems Structure and Systems Behavior*, CreateSpace Independent Publishing Platform, 2017.

[Chec99] Checkland, P., *Systems Thinking, Systems Practice: Includes a 30-Year Retrospective*, 1st Edition, Wiley, 1999.

[Craw15] Crawley, P. et al., *System Architecture: Strategy and Product Development for Complex Systems*, Prentice Hall, 2015.

[Dam06] Dam, S., *DoD Architecture Framework: A Guide to Applying System Engineering to Develop Integrated Executable Architectures*, BookSurge Publishing, 2006.

[Date03] Date, C. J., *An Introduction to Database Systems*, 8th Edition, Addison Wesley, 2003.

[Denn08] Dennis, A. et al., *Systems Analysis and Design*, 4th Edition, Wiley, 2008.

[Dori95] Dori, D., "Object-Process Analysis: Maintaining the Balance between System Structure and Behavior," *Journal of Logic and Computation* 5(2), pp.227-249, 1995.

[Dori02] Dori, D., *Object-Process Methodology: A Holistic Systems Paradigm*, Springer Verlag, New York, 2002.

[Dori16] Dori, D., *Model-Based Systems Engineering with OPM and SysML*, Springer Verlag, New York, 2016.

[Elma10] Elmasri, R., *Fundamentals of Database Systems*, 6th Edition, Addison Wesley, 2010.

[Hoar85] Hoare, C. A. R., *Communicating Sequential Processes*, Prentice-Hall, 1985.

[Jevo12] Jevon, P. et al., *Monitoring the Critically Ill Patient*, 3rd Edition, Wiley-Blackwell, 2012.

[Kend10] Kendall, K. et al., *Systems Analysis and Design*, 8th Edition, Prentice Hall, 2010.

[Maie09] Maier, M. W., *The Art of Systems Architecting*, 3rd Edition, CRC Press, 2009.

[Miln89] Milner, R., *Communication and Concurrency*, Prentice-Hall, 1989.

[Miln99] Milner, R., *Communicating and Mobile Systems: the π-Calculus*, 1st Edition, Cambridge University Press, 1999.

[O'Rou03] O'Rourke, C. et al, *Enterprise Architecture Using the Zachman Framework*, 1st Edition, Course Technology, 2003.

[Pele00] Peleg, M. et al., "The Model Multiplicity Problem: Experimenting with Real-Time Specification Methods". *IEEE Tran. on Software Engineering*. 26 (8), pp. 742–759, 2000.

[Prat00] Pratt, T. W. et al., *Programming Languages: Design and Implementation*, 4th Edition, Prentice Hall 2000.

[Pres09] Pressman, R. S., *Software Engineering: A Practitioner's Approach*, 7th Edition, McGraw-Hill, 2009.

[Putm00] Putman, J. R. et al., *Architecting with RM-ODP*, Prentice-Hall, 2000.

[Rayn09] Raynard, B., *TOGAF The Open Group Architecture Framework 100 Success Secrets*, Emereo Pty Ltd, 2009.

[Roza11] Rozanski, N. et al., *Software Systems Architecture: Working With Stakeholders Using Viewpoints and Perspectives*, 2nd Edition, Addison-Wesley Professional, 2011.

[Rumb91] Rumbaugh, J. et al., *Object-Oriented Modeling and Design*, Prentice-Hall, 1991.

[Seth96] Sethi, R., *Programming Languages: Concepts and Constructs*, 2nd Edition, Addison-Wesley, 1996.

[Sode03] Soderborg, N.R. et al., "OPM-based Definitions and Operational Templates," *Communications of the ACM* 46(10), pp. 67-72, 2003.

[Somm06] Sommerville, I., *Software Engineering*, 8th Edition, Addison-Wesley, 2006.

[Toga08] The Open Group, *TOGAF Version 9 - A Manual (TOGAF Series)*, 9th Edition, Van Haren Publishing, 2008.

[Your99] Yourdon, E., *Death March: The Complete Software Developer's Guide to Surviving Mission Impossible Projects*, Prentice-Hall, 1999.

[Dori95] Dori, D., "Object-Process Analysis: Maintaining the Balance between System Structure and Behavior," *Journal of Logic and Computation* 5(2), pp.227-249, 1995.

[Dori02] Dori, D., *Object-Process Methodology: A Holistic Systems Paradigm*, Springer Verlag, New York, 2002.

[Dori16] Dori, D., *Model-Based Systems Engineering with OPM and SysML*, Springer Verlag, New York, 2016.

[Elma10] Elmasri, R., *Fundamentals of Database Systems*, 6th Edition, Addison Wesley, 2010.

[Hoar85] Hoare, C. A. R., *Communicating Sequential Processes*, Prentice-Hall, 1985.

[Jevo12] Jevon, P. et al., *Monitoring the Critically Ill Patient*, 3rd Edition, Wiley-Blackwell, 2012.

[Kend10] Kendall, K. et al., *Systems Analysis and Design*, 8th Edition, Prentice Hall, 2010.

[Maie09] Maier, M. W., *The Art of Systems Architecting*, 3rd Edition, CRC Press, 2009.

[Miln89] Milner, R., *Communication and Concurrency*, Prentice-Hall, 1989.

[Miln99] Milner, R., *Communicating and Mobile Systems: the π-Calculus*, 1st Edition, Cambridge University Press, 1999.

[O'Rou03] O'Rourke, C. et al, *Enterprise Architecture Using the Zachman Framework*, 1st Edition, Course Technology, 2003.

[Pele00] Peleg, M. et al., "The Model Multiplicity Problem: Experimenting with Real-Time Specification Methods". *IEEE Tran. on Software Engineering.* 26 (8), pp. 742–759, 2000.

[Prat00] Pratt, T. W. et al., *Programming Languages: Design and Implementation*, 4th Edition, Prentice Hall 2000.

[Pres09] Pressman, R. S., *Software Engineering: A Practitioner's Approach*, 7th Edition, McGraw-Hill, 2009.

[Putm00] Putman, J. R. et al., *Architecting with RM-ODP*, Prentice-Hall, 2000.

[Rayn09] Raynard, B., *TOGAF The Open Group Architecture Framework 100 Success Secrets*, Emereo Pty Ltd, 2009.

[Roza11] Rozanski, N. et al., *Software Systems Architecture: Working With Stakeholders Using Viewpoints and Perspectives*, 2nd Edition, Addison-Wesley Professional, 2011.

[Rumb91] Rumbaugh, J. et al., *Object-Oriented Modeling and Design*, Prentice-Hall, 1991.

[Seth96] Sethi, R., *Programming Languages: Concepts and Constructs*, 2nd Edition, Addison-Wesley, 1996.

[Sode03] Soderborg, N.R. et al., "OPM-based Definitions and Operational Templates," *Communications of the ACM* 46(10), pp. 67-72, 2003.

[Somm06] Sommerville, I., *Software Engineering*, 8th Edition, Addison-Wesley, 2006.

[Toga08] The Open Group, *TOGAF Version 9 - A Manual (TOGAF Series)*, 9th Edition, Van Haren Publishing, 2008.

[Your99] Yourdon, E., *Death March: The Complete Software Developer's Guide to Surviving Mission Impossible Projects*, Prentice-Hall, 1999.

www.ingramcontent.com/pod-product-compliance
Lightning Source LLC
Chambersburg PA
CBHW082338220526
45470CB00008B/2553